Oil & Gas Pipelines

Oil & Gas Pipelines

Editor

Deepa Honap

scitus
academics

Oil & Gas Pipelines

Edited by **Deepa Honap**

Printed in 2017

ISBN: 978-1-68117-424-2

Library of Congress Control Number: 2015936539

© 2016 by

SCITUS Academics LLC,
616, Corporate Way, Suite 2, 4766,
Valley Cottage, NY 10989

www.scitusacademics.com

Contents

Tobias Lienen, Anne Kleyböcker, Manuel Brehmer,
Matthias Kraume, Lucie Moeller, Kati Görsch, and
Hilke Würdemann

Preface

A book aiming to describe all phases of oil and gas pipeline design, construction, and operation can only highlight the skills, equipment, and technology required. Pipeline systems in scores of countries around the world differ in purpose, size, complexity, operating environment, regulatory requirements, economic conditions, and design philosophy. Some aspects of pipeline design and operation are based on physical laws. The relationship between pipeline operating pressure and fluid capacity, for instance, is not affected by political boundaries. Describing such relationships is relatively straightforward. But how each company chooses to control its pipeline, or regulations governing operation and construction often can be introduced only by discussing representative situations in a book of this type.

Editor

Study of Agglomeration Characteristics of Hydrate Particles in Oil/Gas Pipelines

Wuchang Wang, Yuxing Li, Haihong Liu, and Pengfei Zhao

College of Pipeline and Civil Engineering, China University of Petroleum (East China), No. 66, The Yangtze River West Road, Economic & Technological Development Zone, Qingdao 266580, China

ABSTRACT

The force acting on hydrate particles is the critical factor to hydrate slurry stability which serves as fundamental basis for slurry flow assurance. A comprehensive analysis of forces acting on the hydrate particles was executed to determine the major agglomeration forces and separation forces, and comparison of forces reveals that the main agglomeration force is capillary force and the main separation force is shear force. Furthermore, four main influencing factors deciding the hydrate particle agglomeration were also analyzed and calculated, which shows contacting angle of capillary bridge is the most important factor for hydrate particles agglomeration, while interface tension of oil and water is the least important one. Some methods must be adopted to change the surface of hydrate agglomerates from hydrophile to lipophilicity so as to control the agglomeration of hydrate particle, which is the significant guarantee for safe flow of oil and gas transporting pipeline with hydrate particles.

INTRODUCTION

Rapid development of deep-sea oil and gas fields brings out the problem of hydrate plugging which is an increasingly severe problem for normal operation of oil pipeline. And the main methods for hydrate plugging prevention include thermodynamic inhibitors and kinetic inhibitors. However, the thermodynamic inhibitors have been proved to be environmentally harmful; thus a new method, cold flow technique, which concentrates on addition of antiagglomerants (AA) or kinetic inhibitors to assure the particles dispersed in continuous phase during flowing other than prevention of hydrate formation, is brought to light and shows great advantages. Forces acting on hydrate particles and agglomeration characteristic of hydrate particles which are crucial for the application of the new technique in engineering field should be studied thoroughly so as to offer instruction for hydrate slurry steady flow.

Based on literatures, force balance model based on hydrate agglomeration force analysis to predict critical agglomeration size has been studied [1–6], and Camargo and Palermo had put forward a force balance model for hydrate agglomerate [7], which was proposed based on force balance between agglomeration force and separation force of hydrate particles and combination with the porous characteristic of hydrate agglomerate, while the agglomeration force in the theoretical method is in need of further study and a popular way to determine value of forces was to measure adhesion force between hydrate particles by means of micromechanical force apparatus (MMF) [8–11]. However, relative researches are far from satisfaction; in this paper, agglomerating forces among hydrate particles influencing factors on agglomeration were analyzed and calculated so as to give some advice for controlling agglomeration and safeguarding the flow of oil and gas pipelines.

MAIN FORCES ON HYDRATE PARTICLES OF WATER-IN-OIL SYSTEM

Main forces on hydrate agglomeration in a flowing system include gravity, buoyancy, Van der Waals force, capillary bridge force, solid bridge force, collision force, and shear force. The forces may be divided into three different categories according to different roles played during agglomeration process [12]: agglomeration forces including Van der Waals force, capillary bridge force, solid bridge force, and electrostatic force, separation forces which include collision force and shear force, and forces that make hydrate agglomerate prone to spin: sliding friction, rolling friction, and so forth, which are not considered in this paper.

Van Der Waals Force

The Van der Waals force between hydrate agglomerate 1 and agglomerate 1 is a basic agglomeration force and can be expressed as follows:

$$F_{\text{VW}} = \frac{A}{12S^2} \frac{d_{a1} d_{a2}}{d_{a1} + d_{a2}}.$$

(1)

Capillary Bridge Force

A liquid bridge, which is stick to the surface of the agglomerates and attracts them with capillary action, can form and be shown in Figure 1 when two hydrate particles are close enough to each other for strongly hydrophilic characters on the hydrate particle surfaces. And capillary bridge force which changes the force character of hydrate agglomerate is generated and can be expressed as follows:

$$F_{cap} = 2\pi R_a \gamma_{ll} \cos \theta.$$

(2)

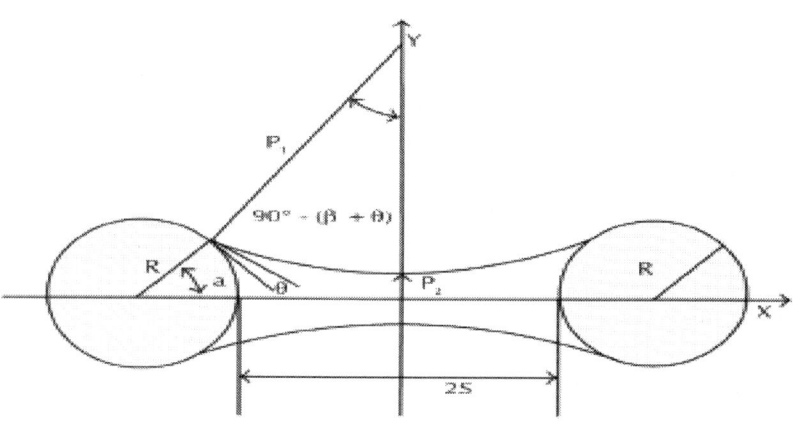

Figure 1: Outline of liquid bridge among hydrate particles.

Solid Bridge Force

Based on theory of hydrate particles contact-inducing agglomeration [16], water contacted with hydrate particles or agglomerates may form hydrates, which can lead to the formation of solid bridge from the liquid bridge and new bigger hydrate agglomerates. There is no well acknowledged formula for solid bridge force calculation, which is usually acquired by experiments. The agglomeration force is not considered in this paper and the formation of solid bridge means that agglomeration has occurred and the agglomerate size has increased. The size of new formed agglomerate instead of the old one should be applied for further calculation of other forces.

Electrostatic Force

The friction and collision among hydrate particles as well as the one between hydrate particles and pipeline wall surface generate electrostatic attraction on the surface of agglomerate; however, the existence of water, which works as electric conductor, diminishes the electrostatic force rapidly [17]; thus the contribution of electrostatic force on agglomeration process is ignored.

Collision Force

For a regular hard sphere collision, the collision force is the key factor to determine if agglomeration happened or not; however, for hydrate particles, the liquid bridge hinders the particle deformation and the collision force is no longer considered as an influencing factor of agglomeration process.

Shear Force

Shear force on the hydrate agglomerates is formed with the flow of the materials in the pipeline always to make the agglomerates flow homogeneously. And shear force is proportional to velocity gradient of fluid media around them:

$$F_{\text{shear}} = 6\pi\mu_0 R_a^2 \gamma.$$

$$(3)$$

Net Gravity Force

The net gravity forces of hydrate agglomerates are defined as the difference between gravity and buoyancy of the hydrate agglomerates and can lead to the deposition of the agglomerates and can be calculated as follows:

$$G = \frac{4}{3}\pi R_a^3 \left(\rho_a - \rho_f \right) g.$$

(4)

DETERMINATION OF MAIN AGGLOMERATION FORCE OF HYDRATE AGGLOMERATE

Basic Parameters

Some basic parameters used in calculating the agglomerate force of hydrate agglomerate are listed in Table 1.

Table 1: Basic parameters [13–15]

Name	Value
Hamaker constant, A	$5 \times 10-21\,J$
Interface tension of oil and water, γ_{ll}	$0.035\,N\cdot m-1$
Viscosity of oil phase, $\mu 0$	$0.135\,Pa\cdot s$
Density of agglomerate, ρ_a	$800\,kg\cdot m-3$

Determination of Main Agglomerate Force of Hydrate Agglomerate

Calculation of hydrate agglomeration forces was carried out based on the previous force analysis, the results of which were shown in Figure 2. Conclusion can be drawn that the value of capillary bridge force is much larger than that of Van der Waals force and net gravity when the hydrate agglomerates have a size from several to several hundreds of micron; thus capillary force is selected as the main agglomeration force of hydrate agglomerate. On the other hand, the main separation force is shear force.

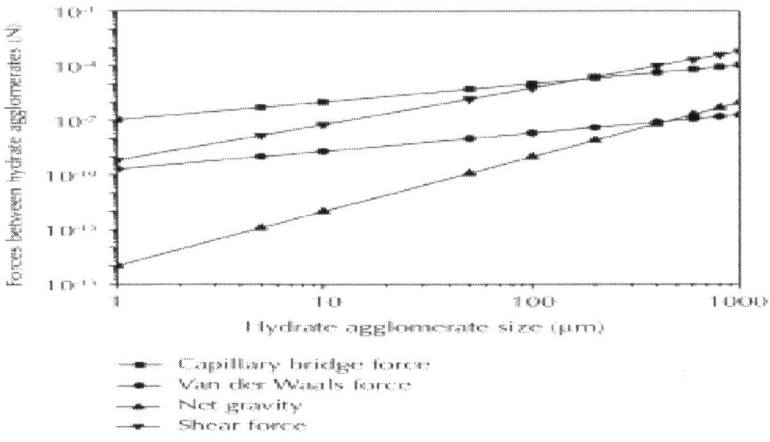

Figure 2: Forces between hydrate agglomerates.

CALCULATION AND VERIFICATION OF THE MAIN FORCES

From the previous force analysis of hydrate agglomerate, we reach a conclusion that the capillary bridge force is much larger than Van der Waals force, so the main agglomeration force of hydrate agglomerate in flowing system is capillary bridge force, and capillary force is used to take place of F_a in the force balance model presented by Camargo and Palermo with a result as follows:

$$\left(\frac{d_A}{d_p}\right)^{3-f} - \frac{\pi\gamma_{il}\cos\theta \cdot \left[1 - (\phi/\phi_{max})(d_A/d_p)^{3-f}\right]^2}{d_p\mu_0\gamma\left[1 - \phi(d_A/d_p)^{3-f}\right]} = 0.$$

(5)

The d_A in the formula actually represents the maximum critical agglomeration size d_{Amax}. In the initial period, hydrate particles are formed and agglomerated with relative small diameters, and then agglomeration goes on with the agglomeration force on the agglomerates larger than separation force. However the agglomeration stops when the diameters of agglomerates reach d_{Amax}. Calculation and verification of hydrate critical agglomeration size with Camargo model were carried out with experimental results from literature (for more details about the experiments, please refer to the literature [11]) and some parameters were decided as follows: fractal dimension $f=2.5$, initial hydrate particle size $d_p=1.5$ μm, viscosity of oil phase $\mu_0=60$ cP, and hydrate volume fraction $\emptyset=0.274$.

Hydrate agglomerate size was calculated by hydrate slurry viscosity, and then agglomeration forces were achieved based on model equation and were shown in Table 2. At the same time, values of Van der Waals force and capillary force were calculated by corresponding calculation formula and listed in Table 2.

Table 2: Agglomerating force calculated by model and the main forces

Shear rate/s−1	Hydrate agglomerate size/μm	Agglomeration force/10−7 N	Van der Waals force/ 10−9 N	Capillary bridge force/ 10−7 N
100	21.44	8.58	4.47	14.5
200	14.76	5.91	3.08	9.97
300	11.64	4.66	2.43	7.86
400	9.69	3.88	2.02	6.54
500	8.4	3.36	1.75	5.68
600	7.71	3.08	1.61	5.21
700	6.96	2.78	1.45	4.70
800	6.46	2.58	1.35	4.36

From the calculation results above, a conclusion can be reached that the capillary force is much closer to agglomeration force with an order of 10^{-7} and more marked than Van der Waals force with an order of 10^{-9}. So the capillary bridge force is consistent with calculated agglomeration force and can be used instead of the agglomerating forces among hydrate particles and agglomerates.

FACTORS INFLUENCING AGGLOMERATION OF HYDRATE PARTICLES

The maximum critical agglomeration size d_{Amax}, which can be calculated by the Camargo and Palermo model with capillary force taking place of F_a and reflects the agglomerating level of hydrate particles, was used as an indicator to study the factors influencing agglomeration of hydrae particles in this paper. And four primary factors including contacting angle of capillary bridge, interface tension of oil and water, viscosity of oil, and shear rate of flow were selected to analyze their influences on hydrate particles.

Influence of Contacting Angle of Capillary Bridge on Agglomeration

Influence of contacting angle of capillary bridge on agglomeration is shown in Figure 3, which indicates that the maximum critical agglomeration size decreases with the increasing of contacting angle of capillary bridge. According to the discussion above, capillary bridge force is the main force deciding the hydrate agglomerates, while the capillary bridge force decreases with the increasing of contacting angles of capillary bridge. Moreover, the surface of hydrate agglomerates changes from hydrophile to lipophilicity with a contacting angle close to 90°, while in the lipophilicity system hydrate agglomerates are apt to spread around and are difficult to agglomerate, which shows that the surface hydrophilic of hydrate agglomerates is a prime factor for the agglomeration of hydrate particles.

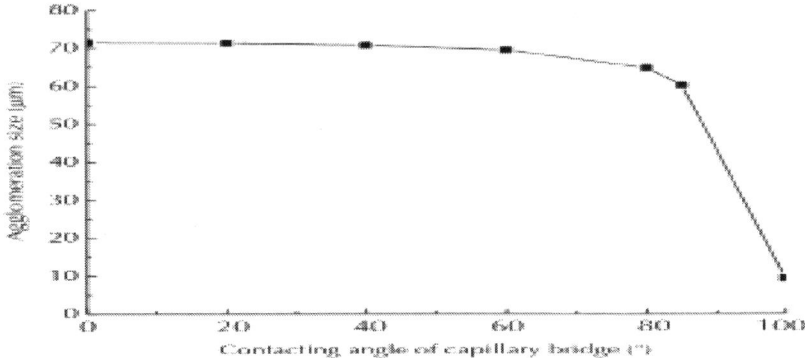

Figure 3: Agglomeration size for different contact angles.

Influence of Interface Tension of Oil and Water on Agglomeration

Influence of interface tensions of oil and water on agglomeration is shown in Figure 4, which indicates that the maximum critical agglomeration size increases almost linearly with the increasing of interface tensions of oil and water. According to the discussion above, capillary bridge force is the main force deciding the hydrate agglomerates, while the capillary bridge force increases linearly with the increasing of interface tensions of oil and water.

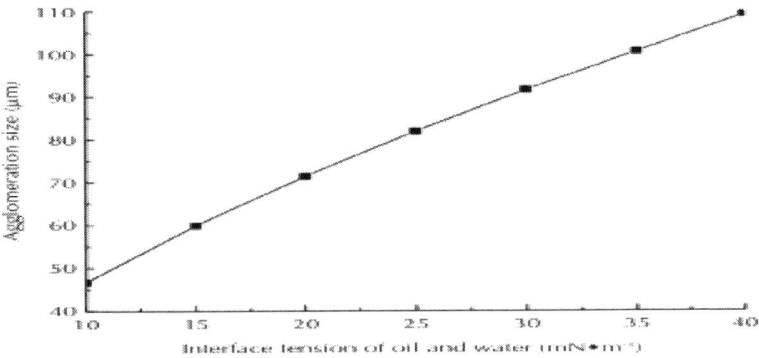

Figure 4: Agglomeration size for different interfacial tensions.

Influence of Oil Viscosity on Agglomeration

Influence of oil viscosity on agglomeration is shown in Figure 5, which indicates that the maximum critical agglomeration size decreases with the increasing of oil viscosity. According to the discussion above, shear force is the main separating force among hydrate agglomerates, while shear force increases with the increasing of oil viscosity.

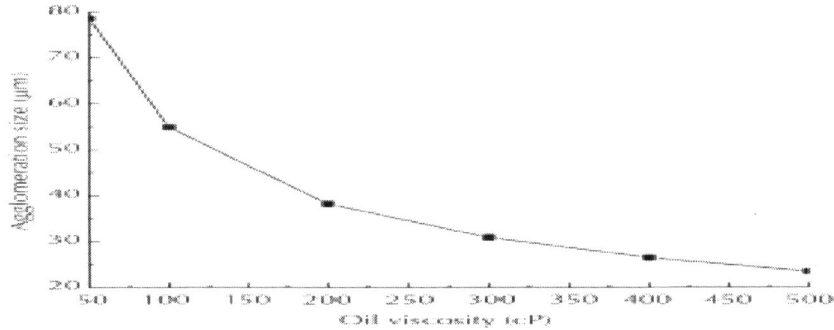

Figure 5: Agglomeration size for different oil viscosity.

Influence of Flow Shearing Rate on Agglomeration

Influence of flow shearing rate on agglomeration is shown in Figure 6, which indicates that the maximum critical agglomeration size decreases with the increasing of flow shearing rate. While all the hydrate agglomerates stop and deposit when the shear rate approaches zero, which means all the hydrates agglomerate to one with a very large size according to calculation, all the agglomerates deposit and form a stationary bed instead of forming one large agglomerate in this situation, which also means the calculation when the shear rate approaches zero will lead to authentic results. According to the discussion above, shear force is the main separating force among hydrate agglomerates, while shear force increases with the increasing of flow shearing rate.

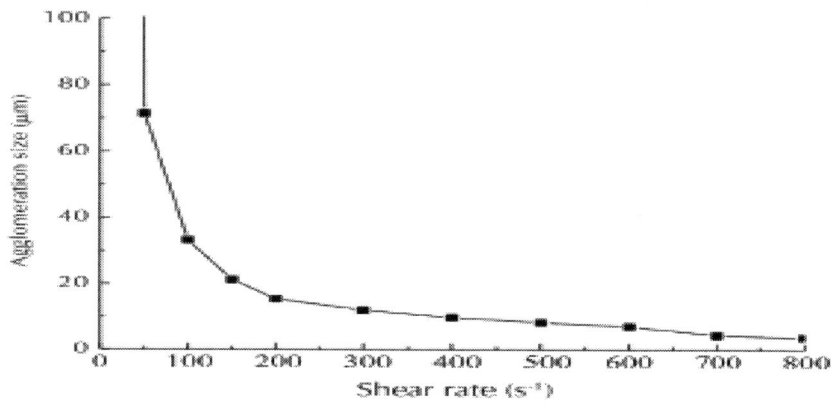

Figure 6: Agglomeration size for different shear rates.

Orthogonal Testing of the Influencing Factors

Four factors discussed above are not isolated with their acting on hydrate agglomeration. Orthogonal testing analysis was used to study the importance of these factors in this paper, in which the maximum critical agglomeration size was used as the evaluating indicator and three levels were selected for each factor as listed in Table 3.

Table 3: Level table for factors of hydrate particle agglomeration

Factors	Levels		
	1	2	3
Contacting angle/°	0	30	90
Interface tension of oil and water/(N/m)	0.01	0.02	0.03
Oil viscosity/(cP)	50	100	200
Shear rate/s−1	100	200	300

Based on the values in Table 3, extreme differences were calculated with orthogonal testing method and listed in Table 4, from which a conclusion can be achieved that contacting angle of capillary bridge is the most important factor for hydrate particles agglomeration, while interface tension of oil and water is the least important one. Some methods must be adopted to change the surface of hydrate agglomerates from hydrophile to lipophilicity so as to control the agglomeration of hydrate particle, which is the significant guarantee for safe flow of oil and gas transporting pipeline with hydrate particles.

Table 4: Range analysis table for factors of hydrate particle agglomeration.

Factors	Contacting angle	Interface tension of oil and water	Oil viscosity	Shear rate
The first mean value	30.43	20.80	33.89	32.61
The second mean value	31.64	22.42	14.87	24.17
The third mean value	4.37	23.22	8.56	9.66
Extreme difference	27.27	2.42	25.33	22.95

CONCLUSIONS

A comprehensive analysis of forces acting on the hydrate particles was executed to determine the major agglomeration force and separation forces, and then four main influencing factors deciding the hydrate particle agglomeration were also analyzed and calculated, based on which the following conclusions are drawn.

- The main agglomeration forces among hydrate particles in a flowing system include capillary bridge force and Van der Waals force, while the main separation force is shear force. Comparison of forces reveals that the main agglomerate force is capillary force and the main separation force is shear force.

- Capillary bridge force is consistent with calculated agglomeration force and can be used to decide the maximum critical agglomeration size among hydrate particles and agglomerates instead of the agglomerating forces.

- Contacting angle of capillary bridge is the most important factor for hydrate particles agglomeration, while interface tension of oil and water is the least important one. Some methods must be adopted to change the surface of hydrate agglomerates from hydrophile to lipophilicity so as to control the agglomeration of hydrate particle.

ACKNOWLEDGMENTS

This study benefits from financial support of National Natural Science Foundation of China (Grant no. 51006120) and Specialized Research Fund for the Doctoral Program of Higher Education (Grant no. 20110133110004).

REFERENCES

1.   T. Zhou, H. Li, and K. Shinohara, "Agglomerating fluidization of group C particles: major factors of coalescence and breakup of agglomerates," Advanced Powder Technology, vol. 17, no. 2, pp. 159–166, 2006.

2.   T. Zhou and H. Li, "Estimation of agglomerate size for cohesive particles during fluidization," Powder Technology, vol. 101, no. 1, pp. 57–62, 1999.

3.   T. Zhou and H. Li, "Force balance modelling for agglomerating fluidization of cohesive particles,"Powder Technology, vol. 111, no. 1-2, pp. 60–65, 2000.

4.   Y. Iwadate and M. Horio, "Prediction of agglomerate sizes in bubbling fluidized beds of group C powders," Powder Technology, vol. 100, no. 2-3, pp. 223–236, 1998.

5.   J. Su and Q.-S. Zhu, "Acting forces and agglomerate sizes of super fine particles in an acoustic fluidized bed," The Chinese Journal of Process Engineering, vol. 10, no. 3, pp. 431–437, 2010.

6.   T. Zhou and H. Li, "The calculation model of agglomerate sizes in fluidized beds of cohesive particles,"Chemical Reaction Engineering and Technology, vol. 15, no. 1, pp. 50–51, 1999.

7.   R. Camargo and T. Palermo, "Rheological properties of hydrate suspensions in an asphaltenic crude oil," in Proceedings of the 4rth International Conference on Gas Hydrates, pp. 880–885, Yokohama, Japan, 2002.

8.   S.-O. Yang, D. M. Kleehammer, Z. Huo, E. D. Sloan, and K. T. Miller, "Temperature dependence of particle-particle adherence forces in ice and clathrate hydrates," Journal of Colloid and Interface Science, vol. 277, no. 2, pp. 335–341, 2004.

9.   C. J. Taylor, L. E. Dieker, K. T. Miller, C. A. Koh, and E. D. Sloan Jr., "Micromechanical adhesion force measurements between tetrahydrofuran hydrate particles," Journal of Colloid and Interface Science, vol. 306, no. 2, pp. 255–261, 2007.

10. L. E. Dieker, Z. M. Aman, N. C. George, A. K. Sum, E. D. Sloan, and C. A. Koh, "Micromechanical adhesion force measurements between hydrate particles in hydrocarbon oils and their modifications,"Energy and Fuels, vol. 23, no. 12, pp. 5966–5971, 2009.

11. Z. M. Aman, G. Aspenes, E. D. Sloan, A. K. Sum, and C. A. Koh, "The effect of chemistry and system conditions on hydrate interparticle adhesion forces toward aggregation and hydrate plug formation," inProceedings of the International Symposium on Oilfield Chemistry, pp. 578–586, The Woodlands, Tex, USA, April 2011.

12. W. Zhang, H. Qi, C. You, and X. Xu, "Mechanical analysis of agglomeration and fragmentation of particles during collisions," Journal of Tsinghua University, vol. 42, no. 12, pp. 1639–1643, 2002.

13. M. R. Anklam, J. D. York, L. Helmerich, and A. Firoozabadi, "Effects of antiagglomerants on the interactions between hydrate particles," American Institute of Chemical Engineers Journal, vol. 54, no. 2, pp. 565–574, 2008.

14. Y. H. Li, w. Yefei, R. Shang, and Z. Fulin, "Influence factors in the interfacial tension between crude oil and water," Journal of Chengde Petroleum College, vol. 8, no. 1, pp. 1–3, 2006.

15. G. Chen, C. Sun, and Q. Ma, Science and Technology of Gas Hydrate, Chemical Industry Press, Beijing, China, 2007.

16. A. Fidel-Dufour, F. Gruy, and J.-M. Herri, "Rheology of methane hydrate slurries during their crystallization in a water in dodecane emulsion under flowing," Chemical Engineering Science, vol. 61, no. 2, pp. 505–515, 2006.

17.   J. Zheng, Kinetic theory of cohesive particles flow and numerical simulation of gas-solid two-phase flows [Dissertation], Harbin Institute of Technology, Harbin, China, 2008.

Chapter 2

Optimal Energy Consumption Analysis of Natural Gas Pipeline

Enbin Liu[1,2], Changjun Li[1,2], and Yi Yang[3]

[1]Southwest Petroleum University, Chengdu 610500, China
[2]CNPC Key Laboratory of Oil & Gas Storage and Transportation, Southwest Petroleum University, Chengdu 610500, China
[3]Beijing Oil and Gas Control Center, Beijing 100191, China

ABSTRACT

There are many compressor stations along long-distance natural gas pipelines. Natural gas can be transported using different boot programs and import pressures, combined with temperature control parameters.

Moreover, different transport methods have correspondingly different energy consumptions. At present, the operating parameters of many pipelines are determined empirically by dispatchers, resulting in high energy consumption. This practice does not abide by energy reduction policies. Therefore, based on a full understanding of the actual needs of pipeline companies, we introduce production unit consumption indicators to establish an objective function for achieving the goal of lowering energy consumption. By using a dynamic programming method for solving the model and preparing calculation software, we can ensure that the solution process is quick and efficient. Using established optimization methods, we analyzed the energy savings for the XQ gas pipeline. By optimizing the boot program, the import station pressure, and the temperature parameters, we achieved the optimal energy consumption. By comparison with the measured energy consumption, the pipeline now has the potential to reduce energy consumption by 11 to 16 percent.

INTRODUCTION

Gas pipelines are the bond that connects gas production and consumption; therefore, their operation must be safe, smooth, and effective. In 1961, a US gas pipeline company collaborated with IBM to simulate and optimize the operation of gas pipelines [1]. This represented the prelude to additional optimal operation research on gas transmission pipelines.

In 1983, Goldberg introduced a genetic algorithm, which was one of the most popular optimization algorithms of the time, to optimize the operation of a natural gas pipeline [2]. The optimal solution of this optimization model considered the minimum energy consumption to be the objective function and promoted research on long-distance pipeline operation optimization using intelligent optimization algorithms. Between 1984 and 1997, many scholars, such as Mantri, Renji, Bhaduri, Anglard, Wilson, Ryan, and Berry et al., continued to improve the operation optimization model of gas transmission pipelines, as well as the methods for obtaining solutions [3–10]. In 1998, Carter took advantage of the dynamic programming algorithm for constructing a steady-state operation optimization model of a gas transmission pipeline [11]. Based on his calculations, he concluded that the dynamic programming algorithm converged faster than the annealing and genetic algorithms. By the end of the 20th century, network simulation models and the optimization of the operation technology for natural gas transmission pipelines had reached maturity. The nonlinear operation optimization model for long-distance gas transmission pipelines (including a discrete variable and objective function for minimum energy consumption) had also been recognized. Since then, researchers have made a sustained effort, taking into consideration the various aspects of the optimization algorithm, to solve the network operation optimization model for a gas transmission pipeline more quickly and effectively. For example, in 2000, Sun and others established a comprehensive pipeline operation optimization expert system [12]. This expert system was capable of detecting the pipeline filling state such that the system could decide the control requirements. It was also able to work out the demand of the corresponding energy consumption. Based on these two steps, a fuzzy model can be used to determine the exact extent to which the compressor should be open. In 2002, Cobos-Zaleta and Rios-Mercado used the equation relaxation and expansion valve method to solve the operation optimization model for a gas pipeline [13]. In 2004, Rusnak et al. used the steady optimization simulator for dynamic optimization analysis of long-distance pipelines, with the goal of simulating the minimal energy consumption [14, 15]. After 2008, Yi et al. studied the problem of steady-state optimization operation of a main gas transmission pipeline network under a determined throughput. In these studies, the optimal rule was adopted based on the minimum energy consumption cost [16–19].

In this paper, we aim to characterize long-distance natural gas pipeline operation management. For a given throughput, with the minimum pipeline operation energy consumption as the goal, the gas pipeline optimal operation model can be established. This model is solved using a dynamic programming method to obtain the best operation scheme and the minimum energy consumption for the natural gas pipeline.

MINIMUM ENERGY CONSUMPTION PREDICTION MODEL OF A NATURAL GAS PIPELINE

Natural gas pipeline systems are complicated. They are composed of pipelines, stations, compressors, fluids, external environmental factors, and other components. Based on the Chinese policy for energy savings and emission reduction and the premise of the transportation quantity plan (intake quantity or delivery quantity), the pipeline operation department must configure each station's compressors and determine the operating parameters for each station to reach the lowest energy consumption for the pipeline system.

To study the minimum energy consumption of a natural gas pipeline system, we need to establish a corresponding mathematical model. A reasonable and accurate mathematical model is the key to obtaining the best results.

The Objective Function

During operation, the pipeline's main energy consumption is from the compressor's drive. Therefore, we established an objective function as the goal for minimum production unit consumption, which is expressed as

$$\min F = \frac{\left(S_p \omega_1 + S_g \omega_2\right)}{T_{ur}},$$

(1)

where F is the production unit consumption of the pipeline in kgce/ (10^7 Nm^3 ·km), Sp is the power consumption in kW·h, S_g is the gas consumption in m^3, ω_1 is the electric coal conversion coefficient based on the Chinese National Standard GB2589-81 of 0.1229 kgce/ (kW·h), ω_2 is the gas coal conversion coefficient based on the Chinese National Standard GB2589-81 of 1.33 kgce/m^3, and T_{ur} is the turnover in 10_7 Nm^3 ·km.

The power consumption Sp can be expressed as follows:

$$S_p = \sum_{i=1}^{n} \frac{N_i t_p}{\eta_{ei}}.$$

(2)

The gas consumption S_g can be expressed as

$$S_g = \sum_{i=1}^{n} \frac{N_i t_p}{\eta_{gi}} ge,$$

(3)

where n is the number of compressors, N_i is the shaft power of the ith compressor in kW, t_i is the running time of the ith compressor in h, η_{ei} is the drive motor efficiency of the ith compressor, η_{gi} is the turbine efficiency of the ith compressor, and g_e is the gas loss rate of the gas turbine in Nm³ / /(kW·h). The turnover T_{ur} can be expressed as

$$T_{ur} = 10^{-4} \sum_{i=1}^{n} Q_i L_i t,$$

(4)

Where Q_i is the volume flow of the ith section of the pipeline in Nm³ /d, L_i is the length of the ith section of the pipeline in km, and t is the delivery time in d.

Optimization Variables

The power of the compressor depends on the compression ratio, flow rate, and temperature. Because the inbound traffic of the compressor station is known, the power of the compressor can be simplified into a function of the pressure ratio and temperature. The compressor inlet and outlet temperatures depend on the compression ratio; therefore, the optimization variables can be converted into the compression ratio and thus can be converted into the outbound pressure. The optimization variables of the optimization model, that is, the outbound pressures and the boot number, can be expressed as

$$X_k = \left(P_{dk}, O_i\right),$$

(5)

Where P_{dk} is the outbound pressure of the kth compressor station and O_i is the boot number of the ith compressor station.

Constraint Condition

To guarantee the safe operation of the pipeline and the devices, both the operation parameters of the pipelines and the operation parameters of the devices must be within the permitted range. Namely, the parameters must be satisfied with a series of constraint conditions.

- **Inlet and Outlet Pressure Constraint**: According to the user's need, there are some requirements for the pressures of the subair node. These are expressed as

$$P_{i\,\min} \leq P_i \leq P_{i\,\max} \quad (i = 1, 2, \ldots, n_S),$$

(6)

Where P_i is the pressure of the ith node in Pa, $P_{i\min}$ is the minimum permissible pressure of the ith node in Pa, and $P_{i\,\max}$ is the maximum allowable pressure of the ith node in Pa.

- **Pipeline Strength Constraints**: To ensure the safe operation of the pipelines, the gas pressure must be less than the maximum allowable operating pressure such that

$$P_k \leq P_{k\,\max} \quad (k = 1, 2, \ldots, n_p),$$

(7)

Where P_k is the pressure of the kth pipe in Pa and $P_{k\,\max}$ is the maximum allowable pressure of the kth pipe in Pa.

- **Compressor Performance Constraints**: The compressor power equation is

$$N = \frac{MH}{\eta},$$

(8)

Where M is the overflow rate of the compressor in kg/s, H is the polytropic head of the compressor, and η is the efficiency of the compressor.

The head curve is calculated according to

$$-H = h_1 S^2 + h_2 SQ + h_3 Q^2,$$

(9)

Where h_1, h_2, and $h3$ are the fitting coefficients of the head curve, S is the speed of the compressor, and Q is the actual overflow rate of the compressor in m³/d.

The efficiency curve is calculated according to

$$\frac{-H}{\eta} = e_1 S^2 + e_2 SQ,$$

(10)

Where e_1 and e_2 are the fitting coefficients of the power curve. The buzz curve is calculated according to

$$Q_{surge} = s_1 + s_2 H,$$

(11)

Where Q_{surge} is the surging flow in m³/d and s_1 and s_2 are the fitting coefficients of the buzz curve.

The stagnation curve is calculated according to

$$Q_{stone} = s_3 + s_4 H,$$

(12)

Where Q_{stone} is the stagnation flow in m³/d and s_3 and s_4 are the fitting coefficients of the stagnation curve.

From (9) to (12) are plotted in the figure, forming a closed area. This area is the operating area of the compressor.

- **Compressor Power Constraints:** The power constraints are represented by

$$N_{min} < N < N_{max},$$

(13)

Where N_{min} is the minimum allowable power of the compressor in MW and N_{max} is the maximum allowable power of the compressor in MW.

- **Compressor Speed Constraints:** The speed constraints are represented by

$$S_{min} < S < S_{max},$$

(14)

Where S_{min} is the minimum speed of the compressor in rpm/min and S_{max} is the maximum speed of the compressor in rpm/min.

- **Compressor Outlet Temperature Constraints:** The temperature constraints are represented by

$$T_H < T_{Hmax},$$

(15)

Where T_{Hmax} is the maximum outlet temperature of the compressor in K.

- **Pipeline Pressure Drop Equation:** The pressure of the pipeline is determined by two factors: the value of the frictional pressure drop and the pressure change due to the elevation change. The calculation of the pressure drop is based on the continuity and momentum equations. Introducing the mass flow rate of the gas [20], we obtain

$$M = \frac{\pi}{4} \sqrt{\frac{\left[P_Q^2 - P_Z^2 \left(1 + a\Delta h \right) \right] D^5}{\lambda ZRTL \left(1 + (a/2L) \right) \sum_{i=1}^{n} \left(h_i - h_{i-1} \right) L_i \right)}},$$

(16)

where M is the flow of the gas through the pipes in kg/s, P_Q is the starting pressure of the pipeline in Pa ($P_Q = P_d$), P_z is the end pressure of the pipeline in Pa ($P_z = P_s$), T is the average of the gas flow temperature in K, $_L$ is the length of the pipeline in m, D is the diameter in m, Δh is the elevation difference between the start and end of the pipeline in m, Z is the gas compressibility (i.e., the pressure computation of the BWRS state equation), and λ is the friction factor.

- **Pipe Temperature Drop Formula**: The pipe temperature drop is calculated according to

$$T = T_0 + (T_Q - T_0) e^{-ax},$$

(17)

Where T is the temperature of length x of the pipeline in K, T_0 is the temperature of the pipeline where it is deeply buried in K, and T_Q is the temperature at the start of the pipeline in K.

- **Pipe Network Node Flow Balance Constraints**: For a natural gas pipeline, in any node, according to the law of conservation of mass, the inflow and outflow of the gas should be 0. In general, for a natural gas pipeline network system with Nn node, the gas flow equilibrium equations of the node can be written as

$$\sum_{\substack{k \in C_i \\ i=1}}^{Nn} \alpha_{ik} M_{ik} + Q_i = 0,$$

(18)

where C_i is the set connected to the ith node element, M_{ik} is the absolute value of element into/out of the node flow connected to the ith node, Q_i is the flow in the node exchange with the outside world (flow into the positive, flow out of the negative), and a_{ik} is the coefficient (when traffic flows in, the K node components are +1 and when traffic flows out, the K node components are −1).

The mathematical model can be written in the standard form for optimization models as

$$\min \quad f(x)$$

$$\text{s.t:} \quad g_i(x) \le 0 \quad (i = 1, 2, \ldots, m),$$

(19)

Where x represents the optimization variables and m is the number of constraints.

METHOD FOR MODELING BASED ON DYNAMIC PROGRAMMING

The gas pipeline branch is simplified to a point. The operation process of the pipeline can be regarded as a multistage process. Thus, we can use a dynamic programming algorithm to distribute the optimal ratio of the compressor stations (i.e., the optimal discharge pressure). Suppose the number of compressor stations is n when establishing the dynamic programming model. Treat the gas transmission process from the compressor station of the $(k - 1)$th to the kth as the kth phase of the correspondence problem. The kth stage of the state variables X_k (corresponding to the starting point of the state) is the discharge pressure $P_{d,-1}$ of the kth station. The phase effect for the kth station energy consumption (i.e., the power, as shown in formula (1)), with respect to the pipeline total energy consumption of the optimization goal, can build the optimized dynamic programming model of the pipeline's compressor station pressure ratio.

The algorithm for solving the model is composed of the following components: "determine the state space," "recursive between stations," "recursive within the station," and "backtracking algorithm."

Determine the State Space

In the dynamic programming algorithm, a certain compressor station out of all of the feasible discharge pressures is the state space. The upper boundary of the state space can give the design pressure of the pipeline. The lower boundary, also called the lowest discharge pressure, is difficult to determine. If it is too large, it will increase the unnecessary computation; however, if it is too small, it may miss the optimal solution. We calculated the lowest discharge pressure for the previous compressor station with the limitations of the lowest discharge pressure of this compressor station.

The compressor with the gas turbine or motor drive performs stepless speed regulation, so that the discharge pressure of the compressor station can be within the scope of feasible continuous change. Thus, we must process the state space to obtain the finite state point. In this paper, the outlet pressure range of each compressor station is divided into 300 points to determine the compression ratio of the space.

When the pipeline is running with low throughput, the station operation plan is always run more economically than with a low compression ratio. This must be taken into consideration for circumstances where the pressure is above the permitted level for one of the compressor stations. By setting each station's entrance pressure as part of the state space, the state transition will not leak.

Recursive between Stations

Recursion between stations is a calculation through which the entrance condition of the next compressor station is determined by the outlet condition of the current compressor station, which mainly involves hydraulic and thermodynamic calculation between stations. On the basis of a certain outlet pressure of the compressor station, (16) and (17) can be used to calculate the pressure and point's parameters are

$Q^1_{d,i-1}$, $P^1_{d,i-1}$, $T^1_{d,i-1}$. The flow should provide the corresponding changes if there is an injection or disengagement point. The final figures for flow, pressure, and temperature are obtained from the inlet operation. In the

end, the optimal index $C^1_{d,i-1}$ corresponding to X^1_i should be recorded as the energy consumption of the inlet operation, which reflects the pipeline's energy consumption under the optimal operation scheme from the beginning to the ith station.

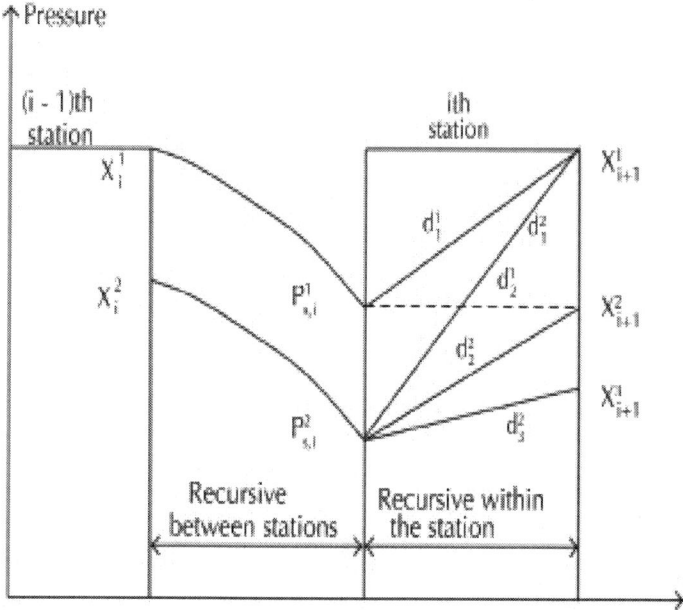

Figure 1: The recursive process.

Recursive within the Station

The recursive within the station gives the outlet station's operation based on the compressor station's inlet operation, which is dominated by the state transfer. For the state before the transfer, in addition to determining the state space, the feasible compression ratio range of compression for every inlet condition should also be obtained, based on the constraint conditions of the decision variables.

Taking the recursive within the station shown in Figure 1 as an example, for X_{i+1}^i, the method of state transition is as follows. Inspect whether the path from $P_{s,i}^1$ to X_{i+1}^i is feasible, which indicates whether d_1^1 gained by X_{i+1}^i divides $P_{s,i}^1$ (station pressure ratio, namely, the decision variables) is in line with the pressure ratio range inlet condition X_{i+1}^i. If not, make the energy consumption of the path a maximum value; otherwise, call for station optimization to obtain the compressor station's optimal scheme under the condition of X_{i+1}^i corresponding to the inlet condition and the station pressure ratio of d_1^j and obtain the energy consumption of the station at the program (if d^1_1 is equal to zero, then so is the energy consumption), namely, the stage effect of the stage. The stage effect and $P^1_{s,}$ of the corresponding inlet condition recorded from the beginning to the energy consumption of this station are added. Then, the total energy consumption from $P_{s,}^1$ to X_{i+1}^i can be obtained. Use the same method to calculate the total energy consumption from $P^2_{s,}$ to X_{i+1}^i. Compared with the former, the smaller one is the X_{i+1}^i state transfer result.

Backtracking Algorithm

After the completion of the recursive within the station, we will obtain all of the total energy costs corresponding to several inlet conditions in the terminal station. To obtain the operation program within the minimum energy consumption limit to meet the terminal station's pressure, backtracking of the whole scheme is required.

Backtracking is performed according to the compression station's inlet and outlet operations recorded in the optimal program to determine the optimal operation scheme of the pipeline. Backtracking starts from the gate station's optimal inlet condition, according to every state transfer's recorded results, to find out every compressor station's outlet condition corresponding to the last station's outlet condition.

OPERATION OPTIMIZATION OF THE XQ GAS PIPELINE

Basic Parameters of the XQ Gas Pipeline

Pipe Parameters

The length of the pipeline is 3840 km, the design capacity is 170 $\times 10^8$ Nm³/year, the design pressure is 10 MPa, and the diameter is

1016×17.5 mm. The elevation and mileage of the XQ gas pipeline are shown in Figure 2. We can see that the elevation change is large, with the highest point at 1900 m and the lowest point at 1 m.

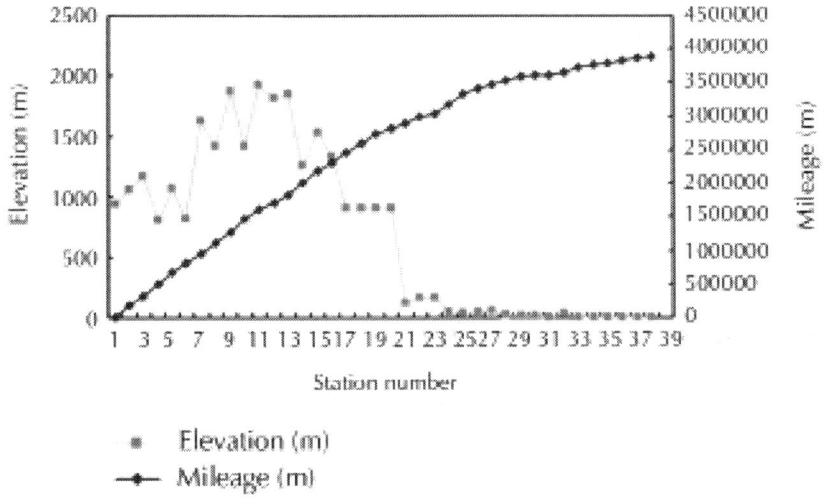

Figure 2: Elevation and mileage of the XQ gas pipeline.

There are 40 stations in the XQ gas pipeline, including 22 compressor stations and 18 distribution stations, as listed in Table 1.

Table 1: Equipment at each station

Station		Compressor		Drive type
Number	Type	Model	Number	
1	Compressor	1	2	Gas
2	Compressor	2	2	Gas
3	Compressor	3	1	Gas
4	Compressor	4	2	Gas
5	Compressor	5	2	Gas
6	Compressor	6	1	Gas
7	Compressor	7	2	Gas
8	Compressor	8	2	Gas

9	Compressor	9	2	Electric
10	Compressor	10	2	Gas
11	Compressor	11	2	Gas
12	Compressor	12	1	Gas
13	Compressor	13	1	Gas
14	Distribution			
15	Compressor	14	1	Gas
16	Compressor	15	2	Gas
17	Compressor	16	1	Gas
18	Distribution			
19	Compressor	17	1	Gas
20	Compressor	18	2	Electric
21	Distribution			
22	Compressor	19	2	Electric
23	Distribution			
24	Compressor	20	2	Electric
25	Distribution			
26	Compressor	21	2	Electric
27	Distribution			
28	Distribution			
29	Compressor	22	2	Gas
30	Distribution			
31	Distribution			
32	Distribution			
33	Distribution			
34	Distribution			
35	Distribution			
36	Distribution			
37	Distribution			
38	Distribution			
39	Distribution			
40	Distribution			

The Compressor Performance Curve

There are two manufacturers for the compressors used in the XQ gas pipeline (GE and RR). Part of the compressor's coefficients for (9)–(12) is shown in Table 2.

Table 2: Coefficients for the equation for the compressor performance curves

Model	H1	H2	H3	e1	e2	S1	S2	S3	S4
1	-0.000282	-0.000393	0.000090	-0.001170	0.000144	4620	0.396	8310	1.44
2	-0.001200	0.000167	0.000045	-0.001470	0.000332	3840	0.145	4910	0.533
3	-0.000403	-0.000348	0.000064	-0.001440	0.000140	5920	0.412	107.00	1.47
4	-0.001200	0.000167	0.000045	-0.001470	0.000332	3840	0.145	4910	0.533
5	-0.000390	-0.001090	0.000334	-0.002180	0.000392	3080	0.145	5970	0.585
6	-0.000640	0.000012	0.000023	-0.000883	0.000141	5010	0.342	8520	1.26
7	-0.000183	-0.001100	0.000314	-0.001990	0.000362	3260	0.173	6270	0.79
8	-0.001190	0.000161	0.000042	-0.001450	0.000317	3610	0.149	4640	0.554
9	-0.000644	-0.000679	0.000252	-0.001790	0.000324	2970	0.165	5340	0.504
10	-0.001190	0.000161	0.000042	-0.001450	0.000317	3610	0.149	4640	0.554

Constraint Conditions

The maximum outbound pressure is 9.8 MPa, while minimum pitted pressure is 5 MPa. The minimum pitted temperature is 15°C, while the maximum outbound temperature is 65°C.

Optimization Research and Analysis

Take the parameters in May 2012 as an example for the optimization calculation. The pitted pressure of the first station is 6.5 MPa and the temperature is 15°C. Each station›s gas transmission capacity is shown in Table 3. There are 5 points for admission and 37 points distributed along the line. Through 50 iterations, the optimum operation is determined, as shown in Table 4, for 23 running compressors. Compressors are connected in parallel at all stations. By means of the energy consumption amount, the energy consumption of the scheme is shown in Table 5. The unit consumption for production is 138.37 kgce/(10^7 Nm^3·km), and the actual measurement of energy consumption is lower by −12.70% compared with the same month, indicating that the pipeline has great potential for saving energy.

Table 3: Transmission capacity, 10^4 Nm^3/d

Station number	Injection volume	Distribution volume
1	3552	0
14	0	291
15	1277	0
21	0	35
22	188	0
23	0	158
24	0	379
25	219	220
26	0	589
27	0	55
28	0	68
29	0	130
30	0	39
31	0	351
32	817	56
33	0	522
34	0	66
35	0	416
36	0	149
37	0	183
38	0	751
39	0	508

Table 4: Optimal operation scheme

Station number	Pitted pressure, MPa	Outbound pressure, MPa	Pitted temperature, °C	Outbound temperature, °C	Compressor boot program
1	6.5	8.43	15	37.28	1 set
2	6.41	9.08	6.61	35.87	2 set
3	7.78	9.75	8.69	27.56	1 set
4	8.5	8.5	6.32	6.32	0 set
5	6.62	9.17	5.09	32.36	2 set
6	8.17	9.78	8.24	23.14	1 set
7	7.96	9.8	6.88	24.09	1 set
8	8.73	8.73	6.73	6.73	0 set
9	6.98	9.8	5.15	33.56	2 set
10	8.54	8.54	6.35	6.35	0 set
11	6.85	9.24	5.15	30.07	2 set
12	8.43	9.8	9.73	22.27	1 set
13	8.81	8.81	7.54	7.54	0 set
15	7.84	9.76	5.18	23.15	1 set
16	7.03	9.7	6.93	33.98	2 set
17	8.05	9.8	11.8	28.32	1 set
19	8.05	9.8	9.41	25.8	1 set
20	7.95	9.74	9.44	26.35	1 set
22	7.68	9.34	8.9	25.18	1 set
24	7.5	9.26	8.36	25.87	1 set
26	7.24	9.03	7.37	25.72	1 set
29	6.29	7.84	5.24	23.39	1 set

Table 5: Energy consumption of the optimal operation scheme

Turnover	$452892.63 \times 107\,\mathrm{Nm^3 \cdot km}$
Gas consumption	$4154.5 \times 104\,\mathrm{Nm^3}$
Gas unit consumption	$91.7\,\mathrm{Nm^3/(107Nm^3 \cdot km)}$
Production unit consumption	$135.4\,\mathrm{kgce/(107Nm^3 \cdot km)}$
Power consumption	$4195 \times 104\,\mathrm{kW \cdot h}$
Total energy consumption	$61314.19\,\mathrm{tce}$
Power unit consumption	$108.9\,\mathrm{kW \cdot h/(107\,Nm^3 \cdot km)}$

Using the same method to optimize the operation for 1–7 months in 2012, the energy consumption optimization results can be obtained. As shown in Table 6, 1–3 months is the gas use peak in the winter. The first station's intake is approximately 4800×10^4 Nm³ per day at full load. The period from 4 to 7 months without heating gas is the low point. The first station's intake is approximately 3500×10^4 Nm³ per day, according to the optimal operation scheme proposed in this paper.

Table 6: XQ1 energy consumption

Month	Production unit consumption, kgce/(10⁷ Nm³·km)			Turnover, 10⁷ Nm³·km	Power consumption, 10⁴ kW·h			Gas consumption, 10⁴ Nm³		
	Optimal value	Measured value	Energy saving rate		Optimal value	Measured value	Deviation	Optimal value	Measured value	Deviation
1	205.1	241.46	−15.07%	521623	5446	5302	2.72%	7540	8980	−16.04%
2	210.1	237.6	−11.56%	496704	4944	5825	−15.12%	7391	8335	−11.33%
3	236.0	283	−16.59%	476826.49	5267	6996	−24.71%	7976	9529	−16.30%
4	147.0	174	−15.53%	452892.63	4449.69	5345	−16.75%	4594	4377	4.96%
5	135.4	158.5	−14.58%	452892.63	4930.3	4195	17.53%	4154.5	5011	−17.09%
6	148.6	171.2	−13.20%	463643.35	5024.59	5194	−3.26%	4716.23	5487	−14.05%
7	157.4	187.3	−15.97%	486394.54	5049.2	4612	9.48%	5289.32	6424	−17.66%

We can acquire the operating parameters through the SCADA systems of the pipeline, including the gas consumption and electricity consumption. Therefore, we can obtain the actual energy of the pipeline in Table 6.

The data in Table 6 is plotted in Figure 3. Compared with the measured values, the production unit consumption can be reduced by approximately 11%~17%. Therefore, the pipeline has great energy-saving potential.

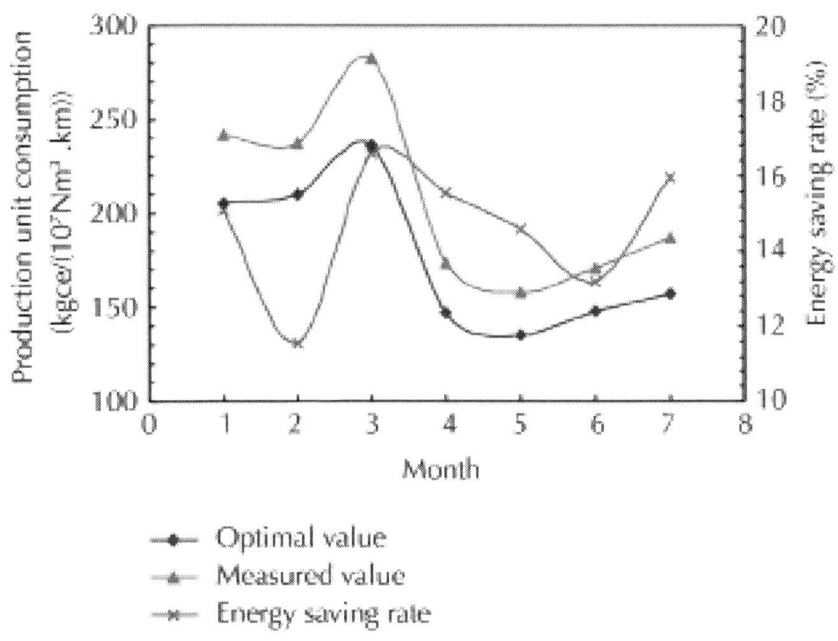

Figure 3: Energy analysis.

CONCLUSIONS

Our conclusions are as follows.

- Based on a full understanding of actual demands of a pipeline company, we introduce production consumption indicators to establish an objective function of the minimum energy consumption of the gas pipeline and use dynamic programming to solve the model quickly and efficiently.

- When setting the constraints, it is necessary to consider the pipeline, station, power equipment, topography, and climate and to simplify these constraints reasonably such that the mathematical model can accurately describe not only the energy consumption of crude oil pipeline but also the convenient mathematical operations.

- According to the dynamic programming method, we compiled the natural gas pipeline running optimization software, which can be used to guide the natural gas pipeline running program analysis and optimize the energy savings. Through the optimization analysis of the XQ nature gas pipeline with the actual working condition, we discovered that the optimal operation scheme can reduce energy consumption by 11%~16%.

ACKNOWLEDGMENTS

This work was supported by the special fund of China's central government for the development of local colleges and universities—the project of National First-Level Discipline in Oil and Gas Engineering, the Scientific Research Cultivate Project of SWPU, the National Natural Science Foundation of China (no. 51174172), and a subproject of the National Science and Technology Major Project of China (no. 2011ZX05054).

REFERENCES

1. L.-G. Li and Z.-H. Zhang, "Operation optimization based on improved pattern search algorithm in gas transmission networks," Journal of China University of Petroleum, vol. 36, no. 4, pp. 139–143, 2012.

2. D. E. Goldberg, Computer-aided gas pipeline operation using genetic algorithms and rule learning [Ph.D. dissertation], University of Michigan, Ann Arbor, Mich, USA, 1983.

3. V. B. Mantri, L. B. Preston, and C. C. Pringle, "Computer program optimizes nature gas pipeline operation," Pipeline Industry, vol. 6, no. 6, pp. 39–45, 1986.

4. Z. Renji, S. Selandari, and H. G. Nicolai, "A system for control and optimum operation of a gas transmission network," in Proceedings of the 19th Annual Pipeline Simulation Interest Group Meeting (PSIG '87), Tulsa, Okla, USA, 1987.

5. M. J. Ryan and P. B. Percell, "Steady state optimization of gas pipeline network operation," in Proceedings of the 19th Annual Pipeline Simulation Interest Group Meeting (PSIG '87), Tulsa, Okla, USA, 1987.

6. T. Williams and G. E. Broadbent, "Optimization in the operation of compressor stations on the moomba to adelaide gas pipeline network," in Proceedings of the 21th Annual Pipeline Simulation Interest Group Meeting (PSIG '89), El Paso, Tex, USA, October 1989.

7. M. A. R. Berry, "Optimizing compressor operation by effective application of performance models," inProceedings of the 23th Annual Pipeline Simulation Interest Group Meeting (PSIG '91), Houston, Tex, USA, 1991.

8. S. Bhaduri and R. K. Talachi, "Optimization of natural gas pipeline design," ASME Petroleun Division, vol. 16, pp. 67–75, 1988.

9. J. G. W. Wilson, "Optimization of large gas networks," in Proceedings of the European Conferce on Mathematics in Industry, The University of Stranthclyde, Glasgow, UK, August 1988.

10. P. Anglard and P. David, "Hierarchical steady state optimization of very large gas pipelines," inProceedings of the 20th Annual Pipeline Simulation Interest Group Meeting (PSIG '88), Toronto, Canada, 1988.

11. R. Carter, "Pipeline optimization: dynamic programming after 30 years," in Proceedings of the 30th Annual Pipeline Simulation Interest Group Meeting (PSIG '98), Denver, Colo, USA, 1998.

12. C. K. Sun, V. Uraikul, C. W. Chan, and P. Tontiwachwuthikul, "An integrated expert system/operations research approach for the optimization of natural gas pipeline operations," Engineering Applications of Artificial Intelligence, vol. 13, no. 4, pp. 465–475, 2000.

13. D. Cobos-Zaleta and R. Z. Rios-Mercado, "A MINLP Model for a Minimizing Fuel Consumption on Natural Gas Pipeline Networks," 2002, http://yalma.fime.uanl.mx/.

14.   D. R. Rusnak and E. P. Ferber, "Automated Pipeline Optimization (APO) for nonimations management & comperssor fuel minimization," in Proceedings of the 36th Annual Pipeline Simulation Interest Group Meeting (PSIG '04), 2004.

15.   J. Stoffregen, K. K. Botros, and D. J. Sennhauser, "Pipeline network optimization application of genetic algorithm methodologies," in Proceedings of the 37th Annual Pipeline Simulation Interest Group Meeting (PSIG '05), San Antonio, Tex, USA, 2005.

16.   Y. Yi and Z. Hongwei, "Research on steady optimization operation of China petroleum main gas transmission pipe network," Shanghai Gas, vol. 2, pp. 10–14, 2008.

17.   C. Borraz-Snchez and R. Z. Ros-Mercado, "Minimizing fuel cost in gas transmission networks by dynamic programming and adaptive discretization," Computers & Industrial Engineering, vol. 1, no. 2, pp. 364–372, 2011.

18.   C. Monteiro, T. Santos, L. A. Fernandez-Jimenez, I. J. Ramirez-Rosado, and M. S. Terreros-Olarte, "Short-term power forecasting model for photovoltaic plants based on historical similarity," Energies, vol. 6, no. 5, pp. 2624–2643, 2013.

19.   B. Tomoiag , M. Chindri , and A. Sumper, "Pareto optimal reconfiguration of power distribution systems using a genetic algorithm based on NSGA-II," Energies, vol. 6, no. 3, pp. 1439–1455, 2013.

20.   C. Li, the Transportation of Natural Gas Pipeline, Oil Industry Press, Beijing, China, 2000.

Chapter 3

Decision Analysis Framework for Risk Management of Crude Oil Pipeline System

Alex W. Dawotola, P. H. A. J. M. van Gelder, and J. K. Vrijling

Hydraulic Engineering Section, Delft University of Technology, Stevinweg 1, 2628 CN Delft, The Netherlands

ABSTRACT

A model is constructed for risk management of crude pipeline subject to rupture on the basis of a methodology that incorporates structured expert judgment and analytic hierarchy process (AHP). The risk model calculates frequency of failure and their probable consequences for different segments of crude pipeline, considering various failure mechanisms. Specifically, structured expert judgment is used to provide frequency of failure assessments for identified failure mechanisms of the pipeline. In addition, AHP approach is utilized to obtain relative failure likelihood for attributes of failure mechanisms with very low probability of occurrence. Finally, the expected cost of failure for a given pipeline segment is estimated by combining its frequency of failure and the consequences of failure, estimated in terms of historical costs of failure from the pipeline operator's database. A real-world case study of a crude pipeline is used to demonstrate the application of the proposed methodology.

INTRODUCTION

Background

Pipelines carry products that are very vital to the sustenance of national economies and remain a reliable means of transporting water, oil, and gas in the world. They are generally perceived as safe with limited number of failures recorded over their service life. However, like any other engineering assets, pipelines are subject to different degrees of failure and degradation. When it occurs, pipeline rupture can be fatal and very disastrous. It is therefore important that they are effectively monitored for optimal operation while reducing failures to acceptable safety limit.

Integrity maintenance of pipelines is a major challenge of service companies, especially those involved in the transmission of oil and gas. Two major factors have been the driving force behind this challenge. These are the need to minimize costs of operation and doing it without compromising on risk. The huge impact of pipeline failure on operational costs has necessitated the development of more effective risk management strategies to help mitigate potential risks. Ideally, most pipeline operators ensure that during design stage, safety provisions are created to comply with theoretical minimum failure rate for the pipeline. Quantitative risk assessment has been a valuable tool to operators in minimizing risk as well as complying with minimum safety requirement for engineering structures. In quantitative risk assessment, an attempt is made to numerically determine the probabilities of rupture caused by various failure mechanisms and the likely consequences of failure in terms of economic loss, human hazards, and degradation of the environment.

Quantitative risk assessment (QRA) of pipeline networks is complex and can sometimes be laborious due to the differences in the system networks. According to Huipeng [1], one approach to simplify QRA process is the use of hierarchical approach. Hierarchical approaches such as fault tree analysis, event tree analysis, and failure mode event analysis have found applications in risk assessment for complex structures as explained in Dhillon and Singh [2]. However, such methodologies are data intensive. The rupture of pipelines occurs in most countries rarely, and as such, the data of failures are often insufficient to carry out a thorough hierarchical approach. Also, when failure data are gathered, the classifications may not cover all the known failure mechanisms and attributes.

In this paper, a systematic approach to risk ranking and risk management of rupture of crude pipelines is presented and applied to a case study. The pipeline is divided into three different segments, and the level of risk for each segment was determined. The proposed methodology involves a combination of two well-known techniques: AHP and Cooke's classical model for expert elicitation in the context of pipeline maintenance decision support. Developed by Saaty [3], AHP fundamentally works by using opinions of experts in developing priorities for alternatives and the criteria used to judge the alternatives in a system. The classical model proposed by Cooke [4] is a structured expert judgment-based approach. The model is able to provide rational probability assessments and has been successfully applied to over forty-five expert elicitation case studies covering both academic and industrial areas by Cooke and Goossens [5].

In the proposed methodology, the classical model was used to obtain frequency of failure due to rupture for an existing crude pipeline system. Five failure mechanisms were considered. These are external interference, corrosion, structural defects, operational errors and other minor failures. Four of the failure mechanisms are further subdivided into attributes as follows:

- external interference (sabotage and mechanical damage)
- corrosion (internal and external corrosion),
- structural defects (construction defect and material defect)

• Operational errors (equipment failure and human error).

Analytic hierarchy process is then used to rank segments of pipeline riskwise by obtaining relative proportion of attributes with respect to the failure mechanisms. The motivation for AHP was due to the realization that experts find it more difficult to estimate the frequency of failure of failure attributes with generally low probability of occurrence. In essence, it was proposed to conduct pairwise ranking of the attributes using AHP. In addition, failure costs for each failure mechanism/attribute was estimated on the basis of historical failure expenditure data obtained from the pipeline company. On the account of the frequency of failure and failure costs, the expected cost of failure due to rupture on each of the pipeline segment is then calculated.

The unique feature of the approach is that two known methodologies are combined to achieve quantitative risk assessment of pipeline assets. One of the benefits of the approach is that the level of subjectivity in AHP is reasonably reduced, since the classical model entails performance-based calibration of the experts. In other words, experts' inputs are utilized on the basis of the consistency of the experts during elicitation process. The risk assessment results include quantitative estimate of frequency of failure instead of relative ranks expected from a stand-alone AHP. The fact that AHP's output are ranks and not probability can be seen as a major setback to its application in risk analysis. By combining quantitative estimates from classical model with relative ranking from AHP, frequencies of failure of pipeline segments can be estimated, taking uncertainty into consideration.

The remainder of the paper is classified into four sections. Section 2 introduces and explains the proposed classical-AHP methodology. Section 3 presents a case study of cross-country crude pipeline to illustrate the proposed methodology. Section 4 applies the frequency of failure and failure costs hitherto obtained in the model to provide risk management philosophy of pipeline segments, and Section 5 draws the conclusion. The risk-based ranking of pipeline segments is valuable to oil and gas companies in prioritizing inspection and maintenance activities of pipelines, ranking causes of failure by severity of impact and in budget allocation to maintenance activities. The results could also prove valuable in arriving at a design, redesign, construction, and monitoring decision for existing and new pipelines.

THE CLASSICAL-AHP METHODOLOGY

In the following sections, a description of analytic hierarchy process and structured expert judgment techniques will be provided in other to provide good background for the application of the proposed methodology in quantitative risk assessment of crude pipelines.

Failure Frequency Calculation Using Structured Expert Judgment (The Classical Model)

The classical model is a formal method for deriving the requisite weights for a linear pool of individual experts. It is a structured expert judgment elicitation approach that involves treating expert judgments as scientific data in a formal decision process. The basic procedures in the classical model are pre-elicitation, elicitation, and post-elicitation. The processes that comprise each step are summarized in Figure 1 below.

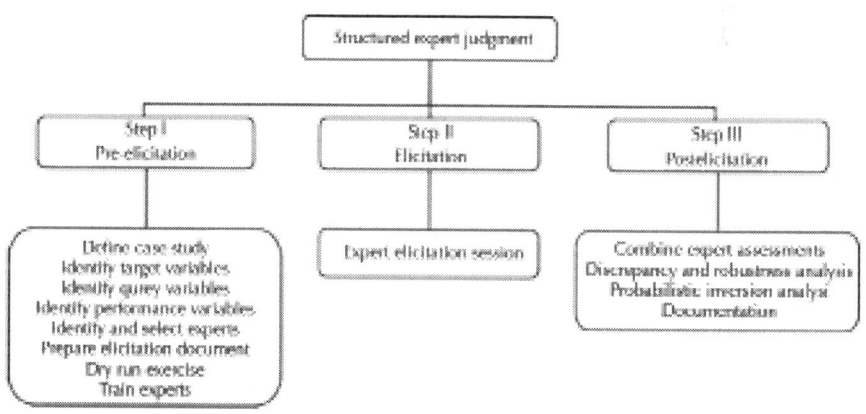

Figure 1: Expert judgment steps.

A major part of the classical model is the requirement that experts should provide information only on quantities which are measurable and familiar to the experts. That is, the quantities for which the experts have to provide information should be verifiable by experiments. The expert's uncertainty distribution is combined using performance-based weighting derived from their responses to the seed variables. The purpose of the seed variables in the model can be classified into three, namely, (i) to quantify experts' performance as subjective probability assessors, (ii) to enable performance-optimized combinations of experts' distributions, and (iii) to evaluate and validate the combination of expert judgments.

The tool used for carrying out structured expert judgment in classical model is the so-called expert calibration software, EXCALIBUR. The software is open access and available through the Risk and Environmental Modeling (REM) group of Delft University of Technology, website: http://risk2.ewi.tudelft.nl. It runs on a windows program that processes parametric and quantile estimates for continuous uncertain quantities into final experts weights on the basis of the classical model. In addition to processing experts structured judgment, EXCALIBUR supports robustness and discrepancy analysis on the results. Robustness analysis shows how sensitive the results are to choice of expert and choice of calibration variables, and discrepancy analysis shows how the experts differ from the decision maker.

In the software, calibration and information scores are combined to derive performance-based weighted combinations of uncertainty distributions of each expert. Information is the degree to which the distribution provided by the expert is concentrated. In the classical model, the amount of concentration is commonly measured by the uniform and log-uniform distributions. Calibration measures the degree to which the actual measured values correspond statistically with the experts assessments. The weights of the classical model are derived from experts' calibration and information scores, as measured on seed variables. Figure 2 is a schematic chart that shows how calibration and information are defined for different experts.

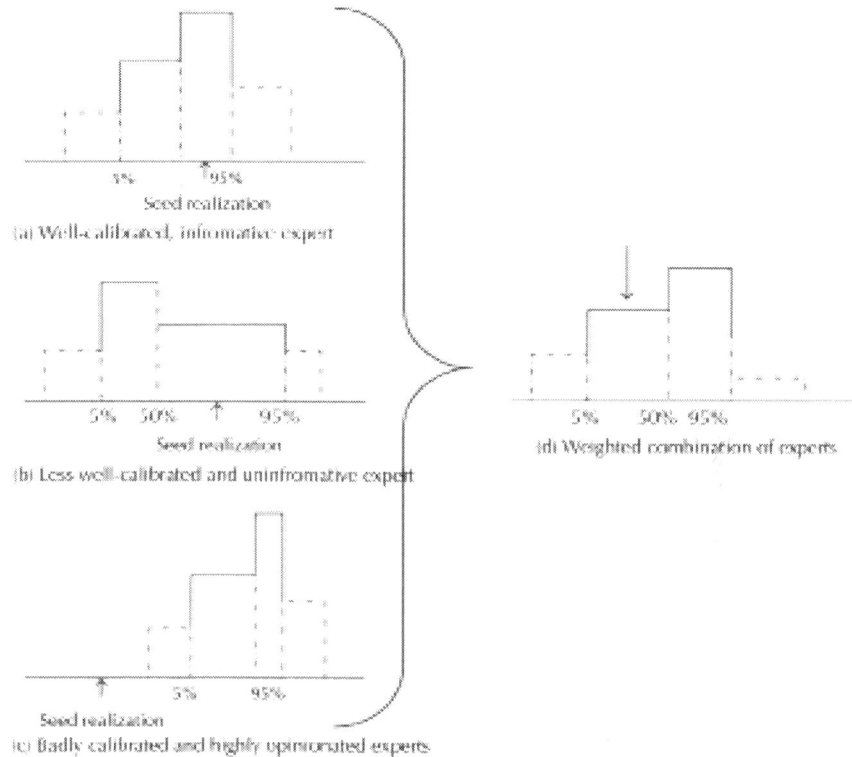

Figure 2: Experts calibration and information: Figures (a–c) show how experts are calibrated on the basis of responses to seed questions at given quantiles. Figure (d) shows the performance-based weighted combination of opinions of experts (a–c) on the target items.

Failure Likelihood Estimate using Analytic Hierarchy Process

Analytic hierarchy process is used in the methodology to rank the attributes of failure mechanisms according to the likelihood of failure for different segments of the pipeline. The outcome is a relative scale which gives a rational basis for risk-based decision making. Analytic hierarchy process has found applications in diverse industries. For example, Quresh and Harrison [6] applied AHP in Riparian revegetation policy selections for a small watershed in Australia. Similarly, Cagno et al. [7] utilised AHP as an elicitation method of expert opinion to determine the a priori distribution of gas pipeline failures, and Dey [8] applied AHP in benchmarking project management practices of Caribbean organizations. The building blocks of analytic hierarchy process are briefly explained below.

Procedures for Analytic Hierarchy Process

The first step of analytic hierarchy process is problem formulation, which involves the ultimate goal for the analysis. In the risk ranking of pipeline, the goal will be selection of pipeline segments with the highest likelihood of rupture due to different failure mechanisms and attributes. Once the goal has been defined, the failure mechanisms are then identified. Thefailure mechanisms are further divided into attributes. The failure mechanisms and attributeswill be in the first and second level hierarchy, respectively, in the AHP value tree.

Secondly, the decision alternatives are selected. The identification of decision alternatives is a very important procedure in analytic hierarchy process. As a matter of fact, the conclusion on the decision alternatives is the outcome of the AHP. For example, the decision maker or an expert could be asked to conduct pairwise assessments of failure mechanisms/attributes of pipeline rupture for a set of pipeline segments. In this case, pipeline segments will be the decision alternatives, and the goal will be to compare these pipeline segments in terms of failure, and to rank them on the basis of the perceived likelihood of rupture.

The next step is the development of hierarchy (value tree). The value tree connects together the goal of the risk assessment, the failure mechanisms and attributes, and the decision variables. In the value tree for risk ranking of crude pipeline, the goal (pipeline selection) is connected to the first level hierarchy (failure mechanisms). The first level hierarchy is then connected to the decision variables (pipeline segments) via the second level hierarchy (attributes).

Thirdly, all necessary information pertaining to the pipeline segments will be collected and recorded. To aid in the classification of the segments, the required features could be divided into physical data, construction data, operational data, inspection data, and failure history. The necessary information on the pipeline/segments should be documented and made available to the experts before pairwise ranking exercise.

Finally, a training session should be organized to familiarize experts with the elicitation procedures. During the elicitation, the experts rank each pair of attribute on the basis of scale proposed by Saaty [3]. Table 1 below gives an explanation of the scale for comparing two attributes. For example, if two criteria are judged to have the same level of risk, the pairwise comparison score will be 1. A score of 9 is given if one criterion is assumed to be extremely stronger than the other. Intermediate judgments of 2, 4, 6, and 8 are selected when a conclusion cannot be reached from the scores of 1, 3, 5, and 7 as defined in Table 1. The responses are consolidated in a preference matrix and synthesized to obtain the weightages.

Table 1: Scale of decision preference for comparing two failure attributes

Judgment	Explanation	Score
Equally	Two attributes have equal likelihood of rupture	1
Moderately	The likelihood of rupture due to one attribute is slightly more than the other attribute	3
Strongly	The likelihood of rupture due to one attribute is strongly more than the other attribute	5
Very strongly	The likelihood of rupture due to one attribute is very strongly more than the other attribute	7
Extremely	The likelihood of rupture due to one attribute is extremely more than the other attribute	9
Intermediate judgment	The intermediate values are used when compromise is needed	2, 4, 6, 8

Consistency Check

AHP provides the possibility of checking the logical consistency of the pairwise matrix by calculating the consistency ratio (CR). The acceptable value for CR is 0.1 maximum, indicating deviations from nonrandom entries of less than an order of magnitude. Factors that affect consistency ratio include homogeneity of attributes of the decision variables, sparseness of the attributes, and the level of knowledge of experts participating in the pairwise ranking of attributes.

Given a weight vector

$$\vec{w} = \begin{bmatrix} w_1 \\ w_2 \\ w_n \end{bmatrix}.$$

(2.1)

Obtained from a decision matrix,

$$A = \begin{bmatrix} a_{11} & a_{12} & a_{1n} \\ a_{21} & a_{22} & a_{2n} \\ a_{n1} & a_{n2} & a_{nn} \end{bmatrix}$$

(2.2)

The consistency of the decision matrix is calculated as follows: multiply matrix A by the weight vector $\vec{\omega}$ to give vector

$$\vec{B} = \vec{A} \cdot \vec{w} = \begin{bmatrix} b_1 \\ b_2 \\ b_n \end{bmatrix},$$

(2.3)

Where

$$b_1 = a_{11}w_1 + a_{12}w_2 + a_{1n}w_n,$$
$$b_2 = a_{21}w_1 + a_{22}w_2 + a_{2n}w_n,$$
$$b_n = a_{n1}w_1 + a_{n2}w_2 + a_{nn}w_n.$$

(2.4)

Divide each element of vector, B with the corresponding element in the weight vector $\vec{\omega}$ to give a new vector

$$c = \begin{bmatrix} b_1/w_1 \\ b_2/w_2 \\ b_n/w_n \end{bmatrix} = \begin{bmatrix} c_1 \\ c_2 \\ c_n \end{bmatrix},$$

(2.5)

λ_{max} is the average of the elements of vector

$$\lambda_{max} = \frac{1}{n}\sum_{i=1}^{n}c_i.$$

(2.6)

Consistency Index is then calculated using,

$$CR = \frac{CI}{RI},$$

(2.7)

Where n is order of the decision matrix and λ_{max} is obtained from (2.6) above.

Using (2.7), consistency ratio is calculated as

$$CR = \frac{CI}{RI},$$

(2.8)

Where RI is the random index and its value is obtained from Table 2 below.

Table 2: Random index table

	3	4	5	6	7	8	9	>9
RI	0.58	0.9	1.12	1.24	1.32	1.41	1.45	1.49

Other measures of consistency have been defined. For example, Mustajoki J and Hämäläinen [9] give a consistency measure (CM) of between 0 to 1 using the multiattribute value theory inherent in their Web-HIPRE software. According to their work, a CM of 0.2 is considered acceptable.

Consistency measure is calculated using

$$CM = \frac{2}{n(n-1)} \sum_{i>j} \frac{\bar{r}(i,j) - \underline{r}(i,j)}{(1 + \bar{r}(i,j))(1 + \underline{r}(i,j))},$$

(2.9)

where $\bar{r}(i,j) = \max a(i,k)a(k,j), k \in \{1,...,n\}$ is the extended bound of the comparison matrix element a(i, j), and r(i, j) is the inverse of r(i, j). CM gives an indication of the size of the extended region formed by the set of local preferences, when $w_i \leq \bar{r}(i,j)w_j$ for all i, j $\in \{1,...,n\}$.

Group Decision Making in AHP

The decision making process in analytic hierarchy process depends on the combination of individual responses of experts to arrive at a group decision. The two big issues in group decision making is how to aggregate individual judgments and how to construct a group choice from individual choices. For programmatic reasons of assignment, it is proposed to aggregate individual judgments using equal weights. Individual expert comparison is combined groupwise by finding the average of individual responses. The average of responses is consistent with the classical model discussed in Section 2.1.

Limitations of Analytic Hierarchy Process

As previously noted, subjectivity limits the outcome of AHP. The presence of subjectivity would introduce uncertainties into the decision making, which could affect the final outcome. In addition, analytic hierarchy process only gives direct qualitative outcomes or relative comparisons. Many researchers such as Cengiz et al. [10], Chang [11], and van Laarhoven and Pedrycz [12] have attempted to fuzzify the results of AHP in other to achieve quantitativeness and reduce subjectivity. However, Saaty and Tran [13] has demonstrated that such approaches are ineffective and capable of creating more uncertainties.

DECISION MODEL APPLICATION

Background Information

The application of the proposed classical-AHP model for risk ranking and assessment is illustrated based on the case study of a crude oil pipeline owned by the Nigerian Petroleum Development Company (NPDC). Some figures of pipeline's failure data have been slightly modified for confidentiality reasons. The pipeline system was commissioned in 1989 and supply crude oil within the south western region of Nigeria. The pipeline is 24 inch in diameter, total length 340 km, with design pressure and operating temperature of 100 bar and 26.8°C, respectively. The material of the pipeline is made from API5LX42 carbon steel, with a concrete type coating. It is basically located onshore but connects a compressor station located offshore.

In the analysis, the entire pipeline is classified into three segments (X1, X2, and X3), in line with its natural stretch. AHP-classical model is utilized to assess the risks related to the pipeline by arranging the segments of pipeline into a hierarchical ranking of risk. The aim of the analysis is to prioritize the most critical segments of pipeline to various failure mechanisms due to rupture. The analysis also takes into consideration the human, environmental, and financial consequences of accidents which may occur in any segment of pipeline.

In order to start the analysis, six pipeline experts from the company were invited and trained on the application of the model. Failure data sheet of each pipeline segment is made available to the experts. The failure data sheet contains information related to pipeline repair history, design parameters, inspection records, and current operating conditions. All the experts are familiar with the pipeline and pipeline segments under study. They participated in both structured expert judgment and AHP-based pairwise ranking of the pipeline segments. The procedure is explained separately below.

Estimation of Failure Frequency Using the Classical Model

Estimation of failure frequencies and uncertainties is carried out on the basis of the classical model. Five failure mechanisms were considered for each pipeline segment, namely, external interference, corrosion, structural defects, operational errors, and other minor failures. The failure mechanisms are actually the target variables in the classical model. In total, twenty eight variables were obtained, considering five target variables for each segment of the pipeline and ten seed variables that are used to calibrate the experts. The seed variables were obtained using generic equipment failure rates from literature and books to calibrate the experts. Initially, the experts were elicitated on the values of the seed variables. Thereafter, each of the experts was required to provide 5%, 50%, and 95% quantiles of the uncertainty distributions for the frequency of failure (in kmyr) by rupture due to the failure mechanisms for segment X1, X2, and X3 of the pipeline.

Expert Calibration

The experts' responses were processed using EXCALIBUR software. The outcome of expert calibration which is based on performance of the "seed" variables are displayed in Table 3. The optimal decision maker (ODM) is also computed. The ODM is obtained as the normalized weighted linear combination of the experts' distributions. In EXCALIBUR, the experts' distributions can be combined using either global weight, item weight, or equal weight. However, in this paper, global weight was used, because it possesses the best calibration and unnormalized weight—which is the combined score of the experts.

Table 3: Results of expert calibration and optimal decision maker

Expert	Calibration	Relative information realization	Unnormalized weight	Normalized weight DM
1	0.036	2.968	0.106	0.248
2	0	3.738	0	0
3	0.001	2.201	0	0
4	0.101	1.906	0.193	0.452
5	0	2.553	0	
6	0.036	3.584	0.128	0.300
Global DM	0.474	1.606	0.761	

Item DM	0.290	1.853	0.537	
Equal DM	0.114	0.989	0.112	

From Table 3, the calibration of the experts reveals that the best experts (in an increasing order) are experts 1, 6, and 4 with normalized weights of 0.248, 0.30, and 0.452 respectively. The other experts (2, 3, and 5) have very low calibration scores, and their individual weights are not considered in the optimal decision maker. Therefore, only experts 1, 6, and 4 form the decision maker. The calibration and information of the optimal decision maker is calculated on the basis of global weight, as discussed before. The outcome confirms the assertion that the ODM calculated on the basis of global weight (calibration = 0.474) possess the best calibration than item weight-based decision maker (calibration = 0.29) and equal weight based decision maker (calibration = 0.11). In addition, it is found that the ODM is better calibrated and its unnormalized weight dominates that of the best experts (1, 4, and 6). However, on the basis of relative information realization, it can be said that the decision maker is less informative than the best experts.

Robustness Analysis

Robustness analysis is performed on the seed variables and the experts. In the robustness analysis, the variables of interest are removed one at a time, and the analysis is repeated until all variables have been covered. The robustness analysis on the experts shown in Table 4indicates that the calibration score for the experts range from 0.474 to 0.55. These scores are well above the calibration score of 0.29 and 0.114 obtained for the item weight DM (item DM) and equal weight DM (equal DM), respectively, in Table 3. Similarly, the robustness of the seed variables is analysed and found to range from 0.405 to 0.731 (Table 5). The initial calibration score obtained for the global DM in Table 3 was 0.474. The analysis confirms the robustness of both the experts and the seed variables, when calibration and information scores of the new decision makers (Tables 4 and 5) are compared with that of the original decision maker (Table3).

Table 4: Robustness analysis of the experts

Excluded expert	Information to background realization	Calibration	Information to original DM realization
1	1.323	0.550	0.285
2	1.606	0.474	0.001
3	1.082	0.474	0.052
4	2.426	0.244	0.824
5	1.199	0.474	0.038
6	1.238	0.474	0.278
None	1.606	0.474	0

Table 5: Robustness analysis of the seed items

Excluded seed variable	Information to background realization	Calibration	Information to original DM realization
1	1.129	0.405	0.549
2	1.569	0.571	0.194
3	1.701	0.405	0.227
4	1.132	0.571	0.167
5	2.452	0.593	0.821
6	1.179	0.731	0.737
7	0.958	0.571	0.593
8	1.626	0.405	0.095
9	1.346	0.571	0.287
10	1.804	0.405	0.133
None	1.606	0.474	0

Resulting Solution

The resulting solution is the combined decision maker's distribution of values assessed by experts that contribute to the ODM. The DM optimization is achieved at a significance level of 0.0358, giving

96.4% acceptable level. The acceptance level is acceptable and the outcome of the structured expert judgment on the frequency of failure of the pipeline due to the identified failure mechanisms for the segments of the pipeline (X1, X2, and X3) is satisfactory. Detailed results of the calculation of failure frequencies are given in Table 6. The 50% uncertainty frequencies of failure for segments X1, X2, and X3 are 2.28E−3 per kmyr, 1.75E−3 per kmyr, and 1.73E−3 per kmyr, respectively.

Table 6: Resulting solution for the decision maker

Item	5%	50%	95%	Failure mechanism
Segment X1				
1-X	0.00025	0.00132	0.00479	Ext. interference
2-X	9.29E−5	0.00045	0.00402	Corrosion
3-X	3.97E−5	0.00022	0.00064	Structural defects
4-X	5.37E−5	0.00016	0.00080	Operational error
5-X	2.37E−5	0.00013	0.00041	Other failures
	4.6E−4	2.28E−3	10.66E−3	Total failure
Segment X2				
1-Y	1.02E−4	0.00114	0.00332	Ext. interference
2-Y	3.20E−5	0.00022	0.00317	Corrosion
3-Y	1.64E−5	0.00016	0.00054	Structural defects
4-Y	2.14E−5	0.00012	0.00059	Operational error
5-Y	1.02E−5	0.00011	0.00033	Other failures
	1.82E−4	1.75E−3	7.95E−3	Total failure
Segment X3				
1-Z	8.20E−5	0.00122	0.00244	Ext. interference
2-Z	2.67E−5	0.00021	0.00241	Corrosion
3-Z	1.36E−5	0.00012	0.00040	Structural defects
4-Z	1.76E−5	0.00020	0.00048	Operational error
5-Z	6.97E−6	0.00008	0.00024	Other failures
	1.47E−4	1.73E−3	5.97E−3	Total failure

The overall failure frequencies compare favourably with results reported in literatures. For example, Little [14] reported a value of

0.42E−3 per kmyr for frequency of failure in Western Europe petroleum pipelines, 0.3E−3 per kmyr for cross country oil pipelines in United Kingdom, and 0.53E−3 per kmyr for total failure of USA Department of Transportation's liquid pipelines. The difference between these values and the frequency of failure obtained for the case study could be due to factors such as difference in location and physical and process properties of the pipelines. These factors have been shown to have significant influence on frequency of failure of pipelines, according to Restrepo et al. [15].

From Table 6, using 50% quantile estimate, it appears that X1 is the most vulnerable among the three pipeline segments, having the highest frequency of failure, followed by X2 and then X3. However, it is interesting to note that X3 has the highest frequency of failure due to operational error. This can be explained partially by the fact that there are more control valves that involve manual operations in X3 compared to X1 and X2.

Relative Estimate of Failure Attributes

In this step, AHP is utilized to determine the likelihood of rupture due to the failure attributes. The six experts that participated in the study were provided with questionnaires that describe features of pipeline segments X1, X2, and X3. The questionnaires were formulated so as to select pipeline segment on the basis of risk of rupture, considering all the failure attributes(sabotage, mechanical damage, internal corrosion, external corrosion, construction defect, material defect, equipment failure and human error, and minor failures).

Construction of Hierarchy

A hierarchy tree of the decision problem is constructed using Web-HIPRE software, version 1.22. The tree (Figure 3) contains information on the goal (selection of pipeline segment), criteria (failure mechanisms) and subcriteria (attributes). The decision alternatives are the three pipeline segments (X1, X2, and X3). The hierarchy tree structure provides the decision makers an overall view of the entire problem through the linking of the decision variables to the overall goal via the attributes and the criteria. The tree aids the decision maker in comparing elements that are on the same level of hierarchy.

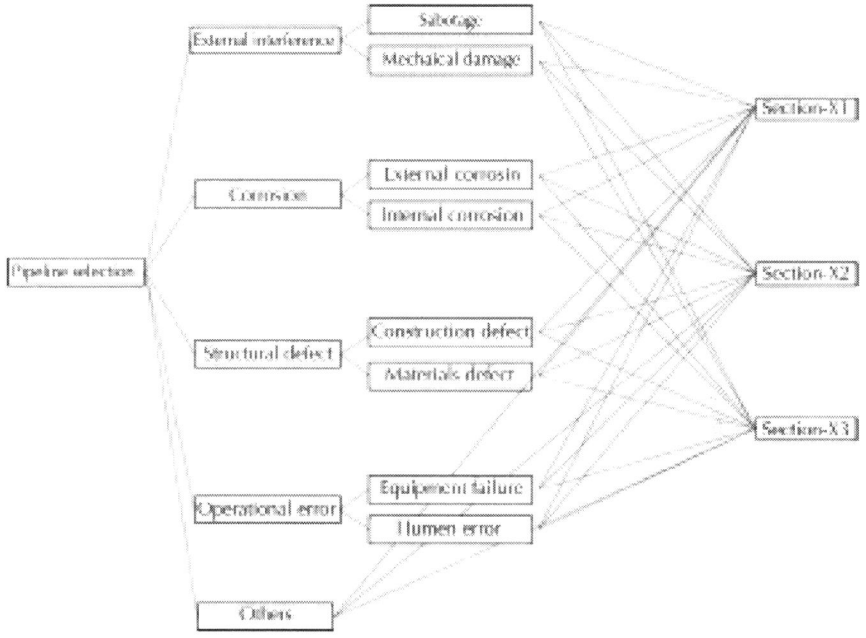

Figure 3: Hierarchical tree for the selection of pipeline segments on the basis of likelihood of rupture.

Results of Pairwise Comparison

Individual expert opinion on the pairwise comparison of the attributes and the pipeline segments were separately collected and analysed using analytic hierarchy process. The outcome of the comparison is a pairwise matrix for the failure likelihood of the pipeline segments on the basis of the judgment of each expert. Initially, the outcome varied from one expert to another until a general session was held in other to establish a common consensus. For all the calculations, the average consistency matrix (CM) obtained from Web-HIPRE software range from 0 to 0.16. Thus, the logical consistency of the elicitation is acceptable.

Results of pairwise comparison of attributes and pipeline segments shown in Table 7 indicate that sabotage contributes 52.5% to the likelihood of pipeline rupture. This was corroborated by the experts. External corrosion, with a percentage of 15.3% has the second highest likelihood of pipeline rupture. The other factors combined accounted for 32.2% of the failure likelihood. The overall failure likelihood for each pipeline segment was synthesized using Web-HIPRE software. The outcome reveals that X1, X2, and X3 have likelihood of failure of 48.8%, 31.6%, and 19.6%, respectively. The conclusion is that X1 is the more prone to rupture, and X3 is the least prone to rupture. The conclusion of AHP analysis is also in agreement with the conclusion from classical model in Table 6.

Table 7: Pairwise ranking of failure criteria and likelihood

Failure mechanisms	Attributes	Likelihood	Pipeline segment		
			X1	X2	X3
External interference	Sabotage	0.525	0.271	0.179	0.076
	Mechanical damage	0.081	0.051	0.019	0.011
Corrosion	External corrosion	0.153	0.093	0.041	0.018
	Internal corrosion	0.061	0.009	0.021	0.031
Structural defects	Construction defect	0.045	0.023	0.014	0.009
	Materials defects	0.021	0.006	0.007	0.008
Operational error	Equipment failure	0.050	0.009	0.018	0.024
	Human error	0.019	0.003	0.007	0.009
Others	Others	0.044	0.023	0.011	0.010
Overall			0.488	0.316	0.196

RISK RANKING AND RISK ASSESSMENT OF PIPELINE

Inspection and Maintenance Strategy

Part of the risk management procedure is to formulate an appropriate inspection and maintenance strategy for pipelines. Broadly speaking, the selection of maintenance strategies for a given failure mechanism depends on a number of factors that include failure attributes, maintenance cost, failure history, level of risk, and acceptability of risk. Table 8 gives some possible strategies for the failure mechanisms and attributes identified for the pipeline under study. However, it should be noted that the selection of a particular inspection technique depends on the experience of pipeline operator.

Table 8: Maintenance strategy for pipeline failures

Failure mechanism	Attributes	Maintenance strategy
External interference	Sabotage	Patrolling
	Mechanical damage	Pipeline marking/improved right of Way
Corrosion	External corrosion	Pipe coating
	Internal corrosion	Intelligent pigging survey
Structural defects	Construction defect	Reconstruction/replacement
	Materials defects	Replacement of pipelines
Operational error	Equipment failure	Replacement of faulty equipment
	Human error	Operator training

Expected Failure Cost

For each pipeline segment, severity of failure was estimated from historical failure costs from database of the pipeline company. The failure costs obtained from the database could not be used directly due to proprietary reasons. The original data was slightly adjusted, and estimates were used in the risk calculations. However, the determination of cost of failure is based on the category of failure. In the Nigerian context, the category of failure in US dollars includes small failure (less than $50,000), medium failure (between $50,000 and up to $200,000), large failure (between $200,000 and $500,000), and catastrophic failure (more than $1 million).

Risk Ranking of Pipeline Segments

In Table 9, pipeline segments X1, X2, and X3 are ranked on the basis of level of risk. The result of frequency of failure (Table 6) shows X1 as the most vulnerable among the three segments, followed by X2 and then X3. However, when failure costs are taken into account and the expected cost of failure is calculated (Table 9) for 50% uncertainty measure of frequency of failure, the trend changed. The system with highest risk remains X1 but X3 now ranked higher based on expected level of risk than X2.

Table 9: Expected failure cost for pipeline segment X1, X2, and X3

Pipeline segment	Frequency of failure (per kmyr)			Failure cost ('$000)	Expected cost of failure ('$000 per kmyr)	Risk ranking
	5%	50%	95%			
X1	4.6E−4	2.28E−3	10.66E−3	5,100	11.6	1
X2	1.82E−4	1.75E−3	7.95E−3	2,095	3.67	3
X3	1.47E−4	1.73E−3	5.97E−3	2,425	4.20	2

In Table 10, the ranking obtained from AHP result in Table 7 is combined with severity of failure to calculate the expected failure cost for each pipeline segment at 50% uncertainty measure of frequency of failure. The expected failure cost calculation shows that the allocation of equal maintenance resources to the three segments will be a less effective maintenance strategy, since they differ in expected cost of failure.

Table 10: Total risk assessments for cross-country crude oil pipeline

Failure mechanism	Pipeline segment	Frequency of failure (per kmyr)			Attributes	Relative rank	Frequency of failure(per kmyr)			Failure cost ($000)	Expected cost of failure ($000 per kmyr)
		5%	50%	95%			5%	50%	95%		
External interference	X1	0.00025	0.00132	0.00479	Sabotage	0.271	2.10E-4	1.11E-3	4.03E-3	2,200	2444
					Mechanical Damage	0.051	3.96E-5	2.09E-4	7.59E-4	1,000	209.1
	X2	1.02E-4	0.00114	0.00332	Sabotage	0.179	9.22E-5	1.03E-3	3.00E-3	800	824.5
					Mechanical Damage	0.019	9.79E-6	1.09E-4	3.19E-4	400	43.8
	X3	8.20E-5	0.00122	0.00244	Sabotage	0.076	7.16E-5	1.57E-3	2.13E-3	1,000	1572.4
					Mechanical Damage	0.011	1.04E-5	2.28E-4	3.09E-4	500	113.8
Corrosion	X1	9.29E-5	0.00045	0.00402	External corrosion	0.093	8.47E-5	4.10E-4	3.67E-3	300	123.1
					Internal corrosion	0.009	8.20E-6	3.97E-5	3.55E-4	200	7.9
	X2	3.20E-5	0.00022	0.00317	External corrosion	0.041	2.12E-5	1.45E-4	2.10E-3	120	17.5
					Internal corrosion	0.021	1.08E-5	7.45E-5	1.07E-3	80	6.0
	X3	2.67E-5	0.00021	0.00241	External corrosion	0.018	9.81E-6	7.71E-5	8.85E-4	120	9.3
					Internal corrosion	0.031	1.69E-5	1.33E-4	1.52E-3	100	13.3

Category	X				Failure mode						
Structural defects	X1	3.97E−5	0.00022	0.00064	Construction defect	0.023	3.15E−5	1.74E−4	5.08E−4	80	14.0
					Material defect	0.006	8.21E−6	4.55E−5	1.32E−4	20	0.9
	X2	1.64E−5	0.00016	0.00054	Construction defect	0.014	1.09E−5	1.07E−4	3.60E−4	30	3.2
					Material defect	0.007	5.47E−6	5.33E−5	1.80E−4	10	0.5
	X3	1.36E−5	0.00012	0.00040	Construction defect	0.009	7.20E−6	1.06E−4	2.12E−4	35	3.7
					Material defect	0.008	6.40E−6	9.41E−5	1.88E−4	15	1.4
Operational error	X1	5.37E−5	0.00016	0.00080	Equipment failure	0.009	4.03E−5	1.20E−4	6.00E−4	800	96.0
					Human error	0.003	1.34E−5	4.00E−5	2.00E−4	400	16.0
	X2	2.14E−5	0.00012	0.00059	Equipment failure	0.018	1.54E−5	8.64E−5	4.25E−4	400	34.6
					Human error	0.007	5.99E−6	3.36E−5	1.65E−4	200	6.7
	X3	1.76E−5	0.00020	0.00048	Equipment failure	0.024	1.28E−5	1.82E−4	3.49E−4	400	72.7
					Human error	0.009	4.80E−6	6.82E−5	1.31E−4	200	13.6
Other failures	X1	2.37E−5	0.00013	0.00041		0.023	2.37E−5	1.30E−4	0.00041	100	13.0
	X2	1.02E−5	0.00011	0.00033	Other Failures	0.011	1.02E−5	1.10E−4	0.00033	55	6.1
	X3	6.97E−6	0.00008	0.00024		0.010	6.97E−6	1.20E−4	0.00024	55	6.6

CONCLUSIONS

A decision-based model has been presented for risk ranking and risk assessment management of crude oil pipelines. The model uses structured expert judgment and analytic hierarchy process to predict the frequency of failure and severity of failure for a given pipeline. The work hopes to contribute to the process of prioritizing transportation pipelines for integrity maintenance on the basis of the results of risk ranking and risk assessment conducted.

The assumption in the AHP model is that each expert would have equal weight in the final decision making. However, the assumption may prevent the decision maker in reaching an optimum conclusion, since equal representation may not always lead to rational consensus. We have been able to demonstrate that an optimum decision making can be achieved with the use of structured expert judgment on the basis of the so-called classical model. The classical model reveals that only three out of the six experts actually contribute to the optimum decision making. In addition, the subjectivity inherent in AHP can be minimized through estimation of uncertainties in the expert elicitation.

The case study revealed some interesting conclusions, which shows that location plays a significant role in pipeline integrity as expected cost of failure vary along pipeline segments. For the case study, external interference is found to be the most important failure criterion, representing over 50% of entire failures. The high likelihood of failure by external interference is due to a somewhat high occurrence of sabotage acts and mechanical damage around the pipeline location. Therefore, increased surveillance along pipeline's right of way would help improve pipeline reliability.

The result also confirms that equal allocation of maintenance resources to pipeline segments may not always be the optimal maintenance decision. For example, in the allocation of maintenance resources for pipeline under study, X1, with the highest expected failure cost should receive more attention than the other segments. In addition, X3 will require more maintenance resources than X2. The maintenance manager will find this approach to be beneficial in formulating the annual inspection and maintenance policy for company's assets. Furthermore, the outcome of the decision analysis could prove useful in formulating individual and societal risk acceptance criteria for regulatory compliance. In general, the accuracy of the severity of failure and the expected cost of failure calculated could be further improved with more pipeline failure data.

ACKNOWLEDGMENTS

The authors would like to acknowledge the management of the Nigerian National Petroleum Company (NNPC) and National Petroleum Development Company (NPDC) for their generous supply of data used in this study. All the experts that participated in this research are also appreciated for their useful contributions.

REFERENCES

1.   L. Huipeng, Hierarchical risk assessment of water supply systems, Ph.D. thesis, Lougborough University, Leicestershire, UK, 2007.

2.   B. S. Dhillon and C. Singh, Engineering Reliability, John Wiley & Sons, New York, NY, USA, 1981.

3.   T. L. Saaty, The Analytic Hierarchy Process, McGraw-Hill, New York, NY, USA, 1980.

4.   R. M. Cooke, Experts in Uncertainty, Environmental Ethics and Science Policy Series, The Clarendon Press Oxford University Press, New York, NY, USA, 1991.

5.   R. M. Cooke and L. L. H. J. Goossens, "TU Delft expert judgment data base," Reliability Engineering and System Safety, vol. 93, no. 5, pp. 657–674, 2008.

6.   M. E. Quresh and S. R. Harrison, "Application of the analytical hierarchy process to Riparian Revegetation Policy options," Small-Scale Forest Economics, Management and Policy, vol. 2, no. 3, pp. 441–458, 2003.

7.   E. Cagno, F. Caron, M. Mancini, and F. Ruggeri, "Using AHP in determining the prior distributions on gas pipeline failures in a robust Bayesian approach," Reliability Engineering and System Safety, vol. 67, no. 3, pp. 275–284, 2000.

8.   P. K. Dey, "Benchmarking project management practices of Caribbean organizations using analytic hierarchy process," Benchmarking, vol. 9, no. 4, pp. 326–356, 2002

9.　J. Mustajoki and R. P. Hämäläinen, "Web-HIPRE: global decision support by value tree and AHP analysis," INFOR, vol. 38, no. 3, pp. 208–220, 2000.

10.　K. Cengiz, T. Ertay, and G. Buyukozkan, "A fuzzy optimization mode f or QFD planning process using analytic network approach," European Journal of Operations Research, vol. 171, no. 2, pp. 390–411, 2006.

11.　D.-Y. Chang, "Applications of the extent analysis method on fuzzy AHP," European Journal of Operational Research, vol. 95, no. 3, pp. 649–655, 1996.

12.　P. J. M. van Laarhoven and W. Pedrycz, "A fuzzy extension of Saaty's priority theory,"Fuzzy Sets and Systems, vol. 11, no. 3, pp. 229–241, 1983.

13.　T. L. Saaty and L. T. Tran, "On the invalidity of fuzzifying numerical judgments in the analytic hierarchy process," Mathematical and Computer Modelling, vol. 46, no. 7-8, pp. 962–975, 2007.

14.　A. D. Little, Risks from Gasoline Pipelines in the UK, CRR 210, Health and Safety Executive, 1999.

15.　C. Restrepo, J. Simonoff, and R. Zimmerman, "Causes, cost consequences, and risk implications of accidents in U.S. hazardous liquid pipeline infrastructure," International Journal of Critical Infrastructures Protection, vol. 2, no. 1-2, pp. 38–50, 2009.

Chapter 4

Feasibility Study on Crack Detection of Pipelines Using Piezoceramic Transducers

Guofeng Du[1,2], Qingzhao Kong[2], Timothy Lai[2], and Gangbing Song[2]

[1]School of Urban Construction, Yangtze University, Jingzhou, Hubei 434023, China

[2]Department of Mechanical Engineering, University of Houston, Houston, TX 77204, USA

ABSTRACT

Damage detection of pipelines is of great significance in terms of safety in the oil and gas industry. Currently, lead zirconate titanates (PZTs) are the most popular piezoceramic materials and show great potential in the applications of structural health monitoring. In this paper, the authors present a feasibility study on the crack detection and severity monitoring of pipelines using PZT transducers. Due to their electromechanical properties, the piezoceramic transducers can be either as an actuator or a sensor to generate or detect the stress wave. The active sensing approach was applied to monitor the crack severity of pipelines. The crack in the stress wave propagation path can be regarded as a stress relief, which reduces the received energy by the sensors. In the test, eight different operating conditions were tested in which one artificial crack was created ranging from 0 mm to 10.5 mm. A wavelet packet-based crack severity index was also built to quantitatively identify the pipeline damage condition at various crack depths.

INTRODUCTION

Pipelines consistently experience complications in service, with some examples being stress corrosion and excessive external forces, which cause the pipelines to form cracks. These cracks, if not detected in a timely fashion, may lead to catastrophic events with severe economic losses and environmental pollution. The study on damage detection of pipelines is of great significance to ensure their safe operation and receives increasing attention in the literature. Methods for pipeline damage detection include the fiber optic sensor based method [1–3], the acoustic emission method [4–6], the ultrasonic method [7–10], the eddy current method [11–13], and piezoelectric impedance method [14–17].

In recent years, the piezoceramic transducer based active-sensing approach has been developed and demonstrated its promises in real-time damage detection and health monitoring of civil infrastructures [18–28]. Due to its advantages of both actuation and sensing capacities, wide bandwidth, fast response, and low cost, piezoceramic based transducers are used in the active sensing approach to structural damage detection. In the active sensing approach, one piezoceramic transducer is used as an actuator to generate the desired wave to propagate through the host structure, and other distributed piezoceramic transducers are used as sensors to detect the wave response. Cracks or damages inside the structure act as a stress relief in the wave propagation path. The amplitude of wave and the transmission energy will decrease due to the existence of cracks or damages. The decrease of the transmission energy can be correlated with the degree of the structural damage. In general, the active sensing approach has the advantages of real time and distributed monitoring.

In this paper, the authors explore the feasibility of applying the active sensing approach to crack detection and crack severity monitoring of pipelines using piezoceramic transducers. The lead zirconate titanate (PZT) type of piezoceramic material is adopted in this paper due to its strong piezoelectric effect. In this research, an experimental setup involving a pipe segment with an artificial crack is fabricated. On the pipe segment, one PZT is used as the actuator in order to generate the swept sine wave signal. Meanwhile, three PZTs are set up at different locations on the pipe as sensors to receive the excitation signal from the actuator. Since all structures have their own unique initial and boundary conditions (including different sensor locations), the severity of structural damage will be assessed by the changes in the damage indexes as monitored by the sensors. With this approach, the severity of the pipeline crack can be monitored by tracing and analyzing the amplitude of the response signal. In addition, the wavelet packet-based crack severity index is implemented to quantify the severity of the crack detected via the active sensing approach.

WAVELET PACKET-BASED CRACK SEVERITY INDEX

When piezoelectric materials are subjected either to a stress or strain, they will generate an electric charge. Similarly, the opposite is also true—when subject to electric charges, piezoelectric materials are able to produce a stress or strain. Due to this special piezoelectric property, PZT transducers can be used interchangeably either as actuators or as sensors. This research takes advantage of these properties in the active sensing approach. One PZT is set up as an actuator to generate a guided electrical signal, while additional distributed PZTs are set up elsewhere to receive this signal. Since the stress wave propagation is highly dependent on the wave path's medium, the characteristics of the received signal can be used as an indicator for the structural health monitoring.

In this research, the basic experimental approach is related to the above principles. The crack in the stress wave propagation path functions as a stress relief. Furthermore, the loss of energy received by the sensors is correlated with the severity of the crack. These phenomena are then quantified with wavelet packet analysis, which is used as a signal-processing tool for analysis. The wavelet transform technique is widely used in engineering structural analysis. For example, the wavelet energy method was used to search the critical incidence of earthquake excitation in multidimensional seismic response of offshore platforms [29]. Wavelet denoising has been used for bridge health monitoring using GPS and the characterization of multipath signals and techniques for their removal by improved particle filtering [30, 31]. In this paper, indicators of damage to the pipeline will be extracted from data using wavelet analysis techniques. Because of the complexity of the tested structure, the frequency response can be observed by a guided swept sine wave input and the energy obtained by wavelet packet analysis from the response is compared to the baseline, thereby increasing the accuracy when judging whether structure damage has occurred. The basic principles of this analysis technique are as follows.

In the proposed health monitoring algorithm, the sensor signal V is decomposed by an n-level wavelet packet decomposition into 2^n signal subsets $\{X_1, X_2, \ldots\ldots, X_{2^n}\}$ and j is the frequency band. The decomposed subset X_j is written as

$$X_j = \left[x_{j,1}\ x_{j,2}, \ldots, x_{j,m}\right], \quad \left(j = 1, 2, \ldots, 2^n\right)$$

(1)

where m is the amount of sampling data. Additionally, the energy of the decomposed signal at time index I can be defined as

$$E_{i,j} = x_{j,1}^2 + x_{j,2}^2 + \cdots x_{j,m}^2$$

(2)

The energy vector at time index i can be defined as

$$E_i = \left[E_{i,1}, E_{i,2}, \ldots, E_{i,2^n}\right]$$

(3)

Based on the calculation of energy vectors (E_i), the crack severity index for the sensor signal at time index i can be expressed as

$$I(i) = \sqrt{\frac{\sum_{j=1}^{2^n} \left(E_{i,j} - E_{1,j}\right)^2}{\sum_{j=1}^{2^n} E_{1,j}^2}}.$$

(4)

Since crack severity may be described by depth, I(i) can be an approximate indicator of the crack size (although it cannot be used to exactly calculate the dimensions). The deeper the crack, the larger the index becomes.

EXPERIMENTAL SETUP AND TEST-ING PROCEDURES

Pipeline Specimen and PZT Locations

One section of a pipeline sample was used in this experiment. The pipeline was constructed from Q235 steel. The dimensions and material properties of the pipeline are shown in Table 1. The outer and inner diameters of the pipeline are 101 mm and 80 mm, respectively.

Table 1: Pipeline dimensions and Q235 steel properties

Steel grade	Density (kg/m3)	Elastic modulus (MPa)	Poisson ratio	Wall thickness (mm)	Pipe length (mm)
Q235	7850	205,000	0.30	10.5	100

Four PZT patches were fixed on the pipeline surface using Epoxy (LOCTITE EPOXY). The locations of PZT patches are shown in Figure 1. It should be noted that PZT-1, PZT-2, and PZT-3 are equidistant from each other along the length of the pipe. Also seen in Figure 1, the location of PZT-4 is rotated 90 degrees counterclockwise from PZT-3. A third point to note is that an artificial crack was cut approximately halfway between PZT-1 and PZT-2. Figure 2 shows the actual specimen with the aforementioned PZT locations and the artificial crack.

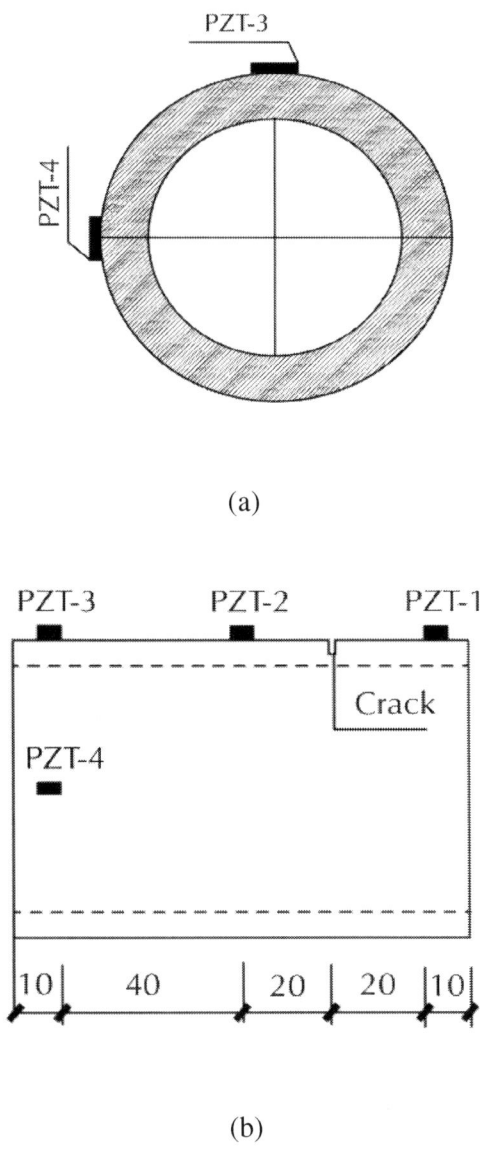

(a)

(b)

Figure 1: Locations of PZT sensors.

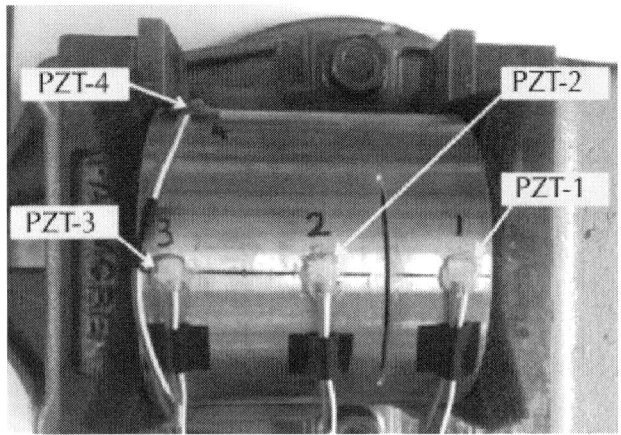

Figure 2: Pipeline specimen with PZT patches.

As mentioned previously, the PZTs display special electromechanical properties, such as density and capacitance. Some of these properties are presented in Table 2. Many factors will influence the detection of pipeline damage, such as piezoelectric ceramic sensor types and properties, the thickness of the bonding layer, the quality of bonding, the material, and size of pipeline. On the other hand, since the proposed method compares the damage indexes of the structure during healthy and damaged states, the above factors will not affect the damage identification results.

Table 2: Main properties of PZT patches

Density (g/cm³)	Dielectric constant	Electromechanical coupling coefficient	Capacitance (nF)	Piezoelectric coefficient (C/N)	Curie temperature (°C)
7.50	1600 ± 10%	0.65	3.77	450	350

During the test, eight operating conditions correlating to different crack depths (0 mm–10.5 mm) were investigated. Table 3 depicts each operating condition with its corresponding crack depth. It should be noted that the crack depth increased by 1.5 mm for each operating condition starting from condition 1 and the crack width was always 1.2 mm.

Table 3: Test operating conditions (OCs)

Operating condition	1	2	3	4	5	6	7	8
Crack depth (mm)	0	1.5	3.0	4.5	6.0	7.5	9.0	10.5

Experimental Setup

In the presented test system, the PZT actuator (PZT-1) is connected with a function generator (Agilent 33120A). The PZT sensors (PZT-2, PZT-3, and PZT 4) are connected with a data acquisition system (NI USB-6363). The sampling rate of the data acquisition system for each channel is 1 Ms/S. The entire experimental setup is shown in Figure 3.

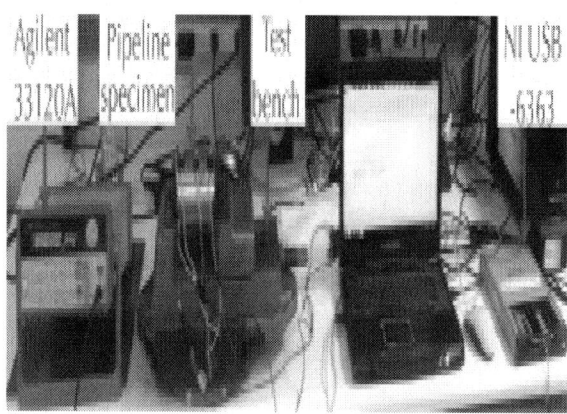

Figure 3: Experimental setup.

Testing Procedures

During the test, PZT-1 was used as an actuator and PZT-2, PZT-3, and PZT-4 were used as sensors. A swept sine wave signal from 60 kHz to 200 kHz was generated by PZT-1, as shown in Figure 4. The amplitude of the excitation signal is 10 V and the period is 2 s. During each operating condition, PZT-1 produced the guided swept sine wave to all the other sensors and the response signal were recorded by the sensors.

Since the pipeline crack was regarded as a stress relief which affected the performance of the stress wave propagation between the actuator and sensors, the sensor signals accurately reflected the pipeline crack severity for each operating condition.

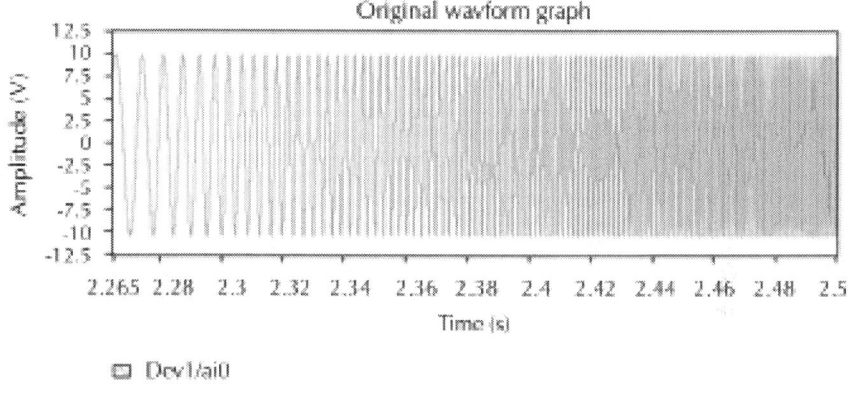

Figure 4: One section of the swept sine wave signal.

EXPERIMENTAL RESULTS AND ANALYSIS

The received signals of PZT-2, PZT-3, and PZT-4 in each operating condition are shown below in Figures 5, 6, and 7. Each signal was subjected to several resonance frequencies within the range of 60 kHz to 200 kHz. Each plot reflects the sensor signal response from one period of the swept sine wave signal, which is equal to 2 seconds in the time domain. From the plots, several resonance peaks can be observed, especially towards the end of the period (i.e., after 1.8 seconds). Through the analysis of these peaks, the same general trend can be identified for each of the PZTs. This trend shows that the amplitudes of the resonance peaks decrease with an increase of the crack depth. This trend indicates that less energy is collected by the sensors with increasing crack depth. Ultimately when the crack is of a depth near 10.5 mm, the entire signal response is extremely weak, which indicates that the crack almost fully blocks the stress wave propagation from the actuator to sensors.

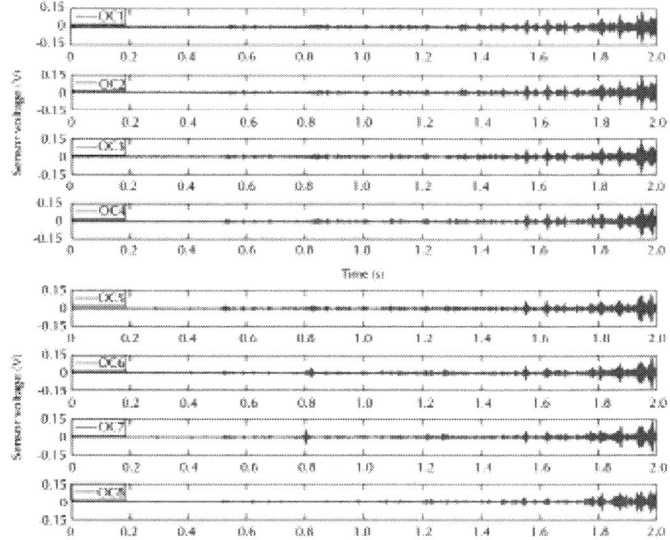

Figure 5: Sensor 2 signal response for each operating condition.

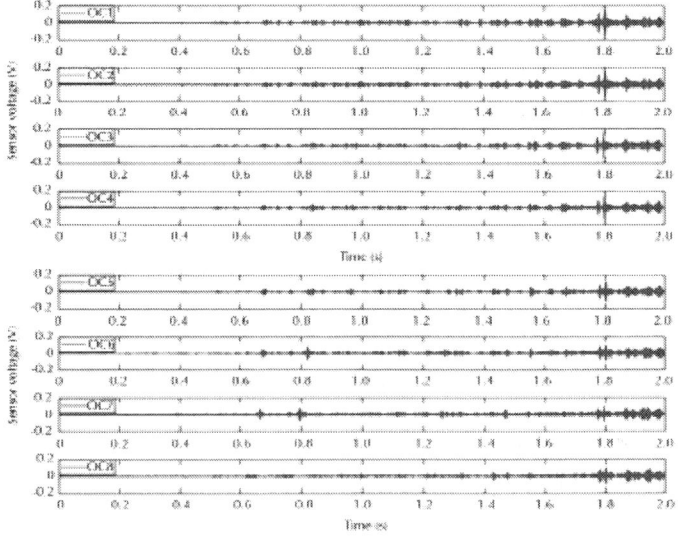

Figure 6: Sensor 3 signal response for each operating condition.

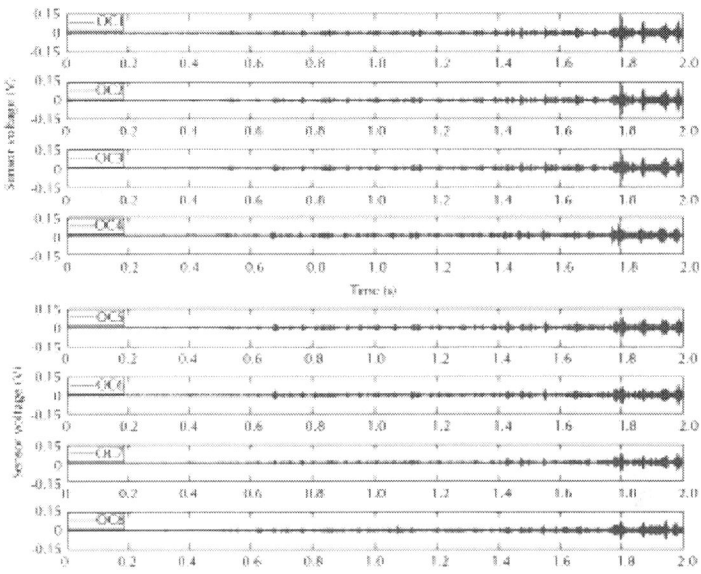

Figure 7: Sensor 4 signal response for each operating condition.

The signals received by sensors are different for the same stress wave propagating through different crack depths. It is from these differences that the method can determine the location and severity of cracks in the structure. As the method only compares two states (healthy versus damaged), differences in materials, size, and so forth across samples/ structures will not affect the performance or requirements of the method. In addition, the vibration response of the structure in a very wide frequency range is calculated by the energy method, and it is more sensitive to minor damage identification.

In order to quantitatively analyze the crack severity on the pipe, the wavelet packet-based crack severity index is developed, as shown in Figure 8. The height of the bars indicates the damage degree collected by the each corresponding sensor. Based on the principle of the crack severity index, 0 is the health status of the structure corresponding to a crack depth of 0 mm (operating condition no. 1). It can be seen that the heights of the bars increase for each incremental operating condition that corresponds to an increase of the crack depth. For Sensor 2, the most distinguishable changes in crack severity index are observed due to the increases in bar height up to the value of 0.4 for operating condition no. 8. This can be attributed to Sensor 2's close proximity to the artificial crack location on the pipe (as seen in Figure 2). The same trend is also observed for Sensors 3 and 4, which confirms that the crack functions as a stress relief in the wave propagation path.

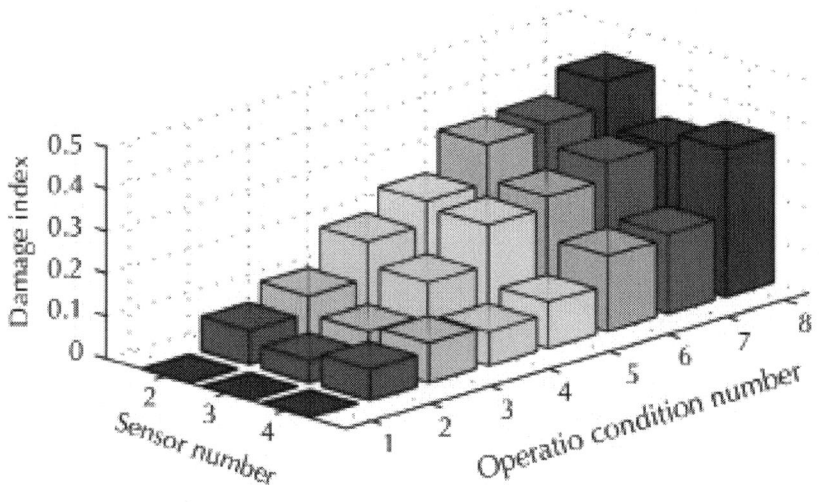

Figure 8: Pipeline crack severity index.

CONCLUSIONS

In this paper, the active sensing based crack severity detection of pipelines was verified. Since the crack functions as a stress relief in the wave propagation path, the signal response of the sensors reports a decreasing trend with the increasing depth of the crack. From the wavelet packet-based crack severity index, the crack severity for each operating condition was quantitatively identified by the values indicated by the heights of the index bars. As the cracks developed, the damage index for all sensors increased. In addition, the energy loss phenomenon directly correlated to the locations of the sensors with respect to the crack and the sensitivity of the sensors. It was then identified that the sensor the closest to the crack was subjected to a largest energy loss. Since the damage index value of the sensors is highly dependent upon sensor locations, the proposed crack severity index presents great potential to locate cracks with distributed sensors.

Compared to the engineering applications for pipelines, the principles and methods of damage detection are identical. To better understand the performance of the active sensing based crack detection system shown in this paper, location detection for multiple cracks on longer pipelines will be studied in future work.

ACKNOWLEDGMENTS

This research reported in this paper was partially supported by The National Natural Science Foundation of China (no. 51378077), the Petro China Innovation Foundation (2011D-5006-0605), and the Scientific Research Project Foundation of Hubei Provincial Department of Education (D20131205).

REFERENCES

1. E. Tapanes, "Fibre optic sensing solutions for real-time pipeline integrity monitoring," Australian Pipeline Industry Association National Convention, 2001.

2. H.-N. Li, D.-S. Li, and G.-B. Song, "Recent applications of fiber optic sensors to health monitoring in civil engineering," Engineering Structures, vol. 26, no. 11, pp. 1647–1657, 2004.

3. S. Z. Yan and L. S. Chyan, "Performance enhancement of BOTDR fiber optic sensor for oil and gas pipeline monitoring," Optical Fiber Technology, vol. 16, no. 2, pp. 100–109, 2010.

4. A. Anastasopoulos, D. Kourousis, and K. Bollas, "Acoustic emission leak detection of liquid filled buried pipeline," Journal of Acoustic Emission, vol. 27, pp. 27–39, 2009.

5. D. Ozevin and J. Harding, "Novel leak localization in pressurized pipeline networks using acoustic emission and geometric connectivity," International Journal of Pressure Vessels and Piping, vol. 92, pp. 63–69, 2012.

6. A. Mostafapour and S. Davoodi, "Analysis of leakage in high pressure pipe using acoustic emission method," Applied Acoustics, vol. 74, no. 3, pp. 335–342, 2013.

7. J. Okamoto, J. C. Adamowskia, M. S. G. Tsuzukia, F. Buiochia,

and C. S. Camerinib, "Autonomous system for oil pipelines inspection," Mechatronics, vol. 9, no. 7, pp. 731–743, 1999.

8.   E. Pan, J. Rogers, S. K. Datta, and A. H. Shah, "Mode selection of guided waves for ultrasonic inspection of gas pipelines with thick coating," Mechanics of Materials, vol. 31, no. 3, pp. 165–174, 1999.

9.   H. Ravanbod, "Application of neuro-fuzzy techniques in oil pipeline ultrasonic nondestructive testing," NDT & E International, vol. 38, no. 8, pp. 643–653, 2005.

10.   B. Hertlein, "Stress wave testing of concrete: a 25-year review and a peek into the future," Construction and Building Materials, vol. 38, pp. 1240–1245, 2013.

11.   D. Vasić, V. Bilas, and D. Ambruš, "Pulsed eddy-current nondestructive testing of ferromagnetic tubes," IEEE Transactions on Instrumentation and Measurement, vol. 53, no. 4, pp. 1289–1294, 2004.

12.   J. B. Nestleroth and R. J. Davis, "Application of eddy currents induced by permanent magnets for pipeline inspection," NDT & E International, vol. 40, no. 1, pp. 77–84, 2007.

13.   R. Keshwani and S. Bhattacharya, "Design and optimization of eddy current sensor for instrumented pipeline inspection gauge," Sensor Review, vol. 28, no. 4, pp. 321–325, 2008.

14.   G. Park, H. H. Cudney, and D. J. Inman, "Feasibility of using impedance-based damage assessment for pipeline structures," Earthquake Engineering and Structural Dynamics, vol. 30, no. 10, pp. 1463–1474, 2001.

15.   D. M. Peairs, G. Park, and D. J. Inman, "Improving accessibility of the impedance-based structural health monitoring method," Journal of Intelligent Material Systems and Structures, vol. 15, no. 2, pp. 129–140, 2004.

16.   K. K. Tseng and L. Wang, "Smart piezoelectric transducers for in situ health monitoring of concrete," Smart Materials and Structures, vol. 13, no. 5, pp. 1017–1024, 2004.

17.   G. F. Du, J. J. Hu, and C. Wan, "The study situation and analysis of oil & gas pipeline health detection," in Proceedings of the International Conference on Pipelines and Trenchless Technology, pp. 551–560, 2012.

18. G. Song, Y. L. Mo, K. Otero, and H. Gu, "Health monitoring and rehabilitation of a concrete structure using intelligent materials," Smart Materials and Structures, vol. 15, no. 2, pp. 309–314, 2006.

19. G. Song, H. Gu, Y. L. Mo, T. T. C. Hsu, and H. Dhonde, "Concrete structural health monitoring using embedded piezoceramic transducers," Smart Materials and Structures, vol. 16, no. 4, pp. 959–968, 2007.

20. L. Jun, "Scattering of harmonic anti-plane shear stress waves by a crack in functionally graded piezoelectric/piezomagnetic materials," Acta Mechanica Solida Sinica, vol. 20, no. 1, pp. 75–86, 2007.

21. G. Song, H. Gu, and Y.-L. Mo, "Smart aggregates: multi-functional sensors for concrete structures—a tutorial and a review," Smart Materials and Structures, vol. 17, no. 3, Article ID 033001, 2008.

22. A. Laskar, H. Gu, Y. L. Mo, and G. Song, "Progressive collapse of a two-story reinforced concrete frame with embedded smart aggregates," Smart Materials and Structures, vol. 18, no. 7, Article ID 075001, 2009.

23. S. Yan, W. Sun, G. Song et al., "Health monitoring of reinforced concrete shear walls using smart aggregates," Smart Materials and Structures, vol. 18, no. 4, Article ID 047001, 2009.

24. P. Li, H. Gu, G. Song, R. Zheng, and Y. L. Mo, "Concrete structural health monitoring using piezoceramicbased wireless sensor networks," Smart Structures and Systems, vol. 6, no. 5-6, pp. 731–748, 2010.

25. H. Gu, Y. Moslehy, D. Sanders, G. Song, and Y. L. Mo, "Multifunctional smart aggregate-based structural health monitoring of circular reinforced concrete columns subjected to seismic excitations,"Smart Materials and Structures, vol. 19, no. 6, Article ID 065026, 2010.

26. Y. Moslehy, H. Gu, A. Belarbi, Y. L. Mo, and G. Song, "Smart aggregate based damage detection of circular RC columns under cyclic combined loading," Smart Materials and Structures, vol. 19, no. 6, Article ID 065021, 2010.

27. W. Liao and J. Wang, "Application of piezoceramic-based sensors to the structural health monitoring of bridge piers," China Civil

Engineering Journal, vol. 45, no. 2, pp. 197–201, 2012.

28. X. Hong, H. Wang, T. Wang, G. Liu, Y. Li, and G. Song, "Dynamic cooperative identification based on synergetics for pipe structural health monitoring with piezoceramic transducers," Smart Materials and Structures, vol. 22, no. 3, pp. 1–13, 2013.

29. H.-N. Li, X.-Y. He, and T.-H. Yi, "Multi-component seismic response analysis of offshore platform by wavelet energy principle," Coastal Engineering, vol. 56, no. 8, pp. 810–830, 2009.

30. T.-H. Yi, H.-N. Li, and M. Gu, "Characterization and extraction of global positioning system multipath signals using an improved particle-filtering algorithm," Measurement Science and Technology, vol. 22, no. 7, Article ID 075101, 2011.

31. T. H. Yi, H. N. Li, and M. Gu, "Wavelet based multi-step filtering method for bridge health monitoring using GPS and accelerometer," Smart Structures and Systems, vol. 11, no. 4, pp. 331–348, 2013.

Corrosion Problems during Oil and Gas Production and its Mitigation

Lekan Taofeek Popoola[1], Alhaji Shehu Grema[2],
Ganiyu Kayode Latinwo[3], Babagana Gutti[2], and
Adebori Saheed Balogun[4]

[1]Department of Petroleum and Chemical Engineering, Afe Babalola
University, Ado-Ekiti, Ekiti State +234, Nigeria

[2]Department of Chemical Engineering, University of Maiduguri,
Maiduguri, Borno State +234, Nigeria

[3]Department of Chemical Engineering, Ladoke Akintola University of
Technology, Ogbomoso, Oyo State +234, Nigeria

[4]Deltaafrik Engineering Limited, Plot 1637, Adetokunbo Ademola,
Victoria Island, Lagos State +234, Nigeria

ABSTRACT

In order to ensure smooth and uninterrupted flow of oil and gas to the end users, it is imperative for the field operators, pipeline engineers, and designers to be corrosion conscious as the lines and their component fittings would undergo material degradations due to corrosion. This paper gives a comprehensive review of corrosion problems during oil and gas production and its mitigation. The chemistry of corrosion mechanism had been examined with the various types of corrosion and associated corroding agents in the oil and gas industry. Factors affecting each of the various forms of corrosion were also presented. Ways of mitigating this menace with current technology of low costs had been discussed. It was noticed that the principles of corrosion must be understood in order to effectively select materials and to design, fabricate, and utilize metal structures for the optimum economic life of facilities and safety in oil and gas operations. Also, oil and gas materials last longer when both inhibitors and protective coatings are used together than when only batch inhibition was used. However, it is recommended that consultations with process, operations, materials, and corrosion engineers are necessary in the fitness of things to save billions of dollars wasted on corrosion in the oil and gas industries.

REVIEW

Introduction

Corrosion is the destructive attack of a material by reaction with its environment [1] and a natural potential hazard associated with oil and gas production and transportation facilities [2]. Almost any aqueous environment can promote corrosion, which occurs under numerous complex conditions in oil and gas production, processing, and pipeline systems [3]. This process is composed of three elements: an anode, a cathode, and an electrolyte. The anode is the site of the corroding metal, the electrolyte is the corrosive medium that enables the transfer of electrons from the anode to the cathode, and the cathode forms the electrical conductor in the cell that is not consumed in the corrosion process [4]. Crude oil and natural gas can carry various high-impurity products which are inherently corrosive. In the case of oil and gas wells and pipelines, such highly corrosive media are carbon dioxide (CO_2), hydrogen sulfide (H_2S), and free water [5]. Continual extraction of CO_2, H_2S, and free water through oil and gas components can over time make the internal surfaces of these components to suffer from corrosion effects. The lines and the component fittings of the lines would undergo material degradations with the varying conditions of the well due to changes in fluid compositions, souring of wells over the period, and changes in operating conditions of the pressures and temperatures. This material degradation results in the loss of mechanical properties like strength, ductility, impact strength, and so on. This leads to loss of materials, reduction in thickness, and at times ultimate failure. A point will be reached where the component may completely break down and the assembly will need to be replaced while production is stopped. The serious consequences of the corrosion process have become a problem of worldwide significance [1].

Corrosion in the modern society is one of the outstanding challenging problems in the industry. Most industrial designs can never be made without taking into consideration the effect of corrosion on the life span of the equipment. Recent industrial catastrophes have it that many industries have lost several billions of dollars as a result of corrosion. Reports around the world have confirmed that some oil companies had their pipeline ruptured due to corrosion and that oil spillages are experienced which no doubt created environmental pollution; in addition, resources are lost in cleaning up this environmental mess, and finally, large-scale ecological damage resulted from corrosion effects [6]. The possibility of occurrence of corrosion in an industrial plant has been posing a lot of concern to petroleum, chemical, and mechanical engineers and chemists. It is now known that corrosion can have some effects on the chemistry of a chosen process, and the product of corrosion can affect reaction and purity of the reaction products.

Many catastrophic incidences resulting from corrosion failure had been historically recorded. On 28 April 1988, a 19-year-old Boeing 737 aircraft, operated by Aloha, lost a major portion of the upper fuselage near the front of the plane due to corrosion damage, in full flight at 24,000 ft [7, 8]. Miraculously, the pilot managed to land the plane on the island of Maui, Hawaii, but one flight attendant died and several passengers sustained serious injuries. Also, the Statue of Liberty which was officially inaugurated on 28 October 1866, on Bedloe's Island, in the New York harbor had undergone severe galvanic corrosion after which remedial measures were taken. The design of the statue rises more than 91 m into the air. Another example of corrosion damage with shared responsibilities was the sewer explosion that killed over 200 people in Guadalajara, Mexico in April 1992 [9]. Besides the fatalities, the series of blasts damaged 1,600 buildings and injured 1,500 people. Damage costs were estimated at 75 million US dollars [10]. The sewer explosion was traced to the installation of a water pipe by a contractor several years before the explosion that leaked water on a gasoline line laying underneath. The subsequent corrosion of the gasoline pipeline, in turn, caused leakage of gasoline into the sewers. The Mexican attorney general sought negligent homicide charges against four officials of Pemex, the government-owned oil company. Also cited were three representatives of the regional sewer system and the city's mayor. Thus, corrosion should be given attention and adequate measures should be taken to curb it as our lives are being endangered in this serious problem.

The costs attributed to corrosion damages of all kinds have been estimated to be of the order of 3% to 5% of industrialized countries' gross national product [11]. The total annual cost of corrosion in the oil and gas production industry is estimated to be $1.372 billion, broken down into $589 million in surface pipeline and facility costs, $463 million annually in downhole tubing expenses, and another $320 million in capital expenditures related to corrosion [12]. Corrosion costs the oil and gas industry tens of billions of dollars in lost income and treatment costs every year [3]. Corrosion costs US industries alone an estimated $170 billion a year in which the oil and gas industry takes more than half of these costs [13]. Internal corrosion in wells and pipelines is influenced by temperature, CO_2 and H_2S content, water chemistry, flow velocity, and surface condition of the steel [14]. Having a greatly reduced corrosion rate (mm/year) can dramatically increase component life, which leads to much greater benefits such as reduced maintenance costs. Currently, many components used for oil and gas extraction are made from carbon steel-based alloys. Now, organizations are looking to move away from these types of alloys to a more corrosion-resistant alloy at a much higher cost. The problem of corrosion is a challenge to the whole world and must be greatly tackled.

Corrosion Types and Associated Agents in the Oil and Gas Industry

The most common form of corrosion in the oil and gas industry occurs when steel comes in contact with an aqueous environment and rusts [4]. When metal is exposed to a corrosive solution (the electrolyte), the metal atoms at the anode site lose electrons, and these electrons are then absorbed by other metal atoms at the cathode site. The cathode, in contact with the anode via the electrolyte, conducts this exchange in an attempt to balance their positive and negative charges. Positively charged ions are released into the electrolyte capable of bonding with other groups of atoms that are negatively charged. This anodic reaction for iron and steel is

$$Fe \rightarrow Fe^{2+} + 2e^-.$$
(1)

After the metal atoms at the anode site release electrons, there are four common cathode reactions [15]:

$O_2 + 4\,H^+ + 4\,e^- \rightarrow 2H^2O$ (oxygen reduction in acidic solution)   (2)

$1/2\,O_2 + H_2O + 2e^- \rightarrow 2OH^-$ (oxygen reduction in neutral or basic solution)   (3)

$2H^+ + 2e^- \rightarrow H_2$ (hydrogen evolution from acidic solution)   (4)

$2H_2O + 2e^- \rightarrow H_2 + 2OH^-$ (hydrogen evolution from neutral water)   (5)

In the oil and gas industry, carbon dioxide (CO_2) and hydrogen sulfide (H_2S) are commonly present, and water is their catalyst for corrosion. When water combines with CO_2 and H_2S, the environments form the following reactions [16]:

H_2CO_3 Reaction: $Fe + H_2CO_3 \rightarrow FeCO_3 + H_2$ (6)

H_2S Reaction: $Fe + H_2S + H_2O \rightarrow FeS + 2H:$ (7)

There may be a combination of the above two reactions if both gases are present. These resulting molecules either attach themselves to the cathode or are released into the electrolyte and the corrosion process continues. Figure 1 is the diagrammatic representation of the corrosion process.

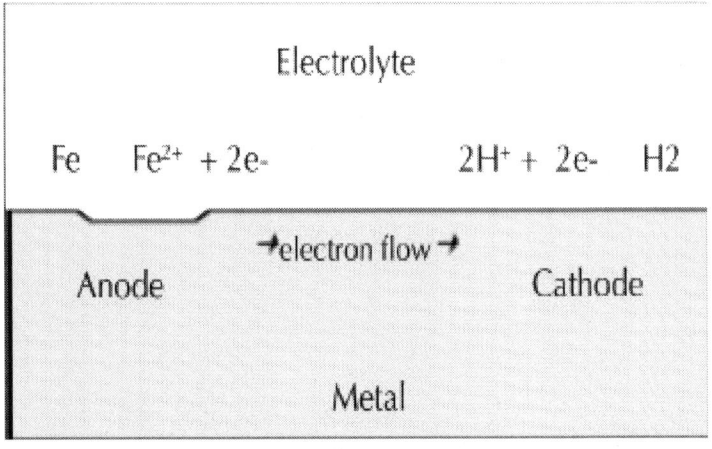

Figure 1: Corrosion process.

It is a great challenge to classify the types of corrosion in the oil and gas industry in a uniform way. One can divide the corrosion on the basis of appearance of corrosion damage, mechanism of attack, industry section, and preventive methods. There are many types and causes of corrosion. The mechanism present in a given piping system varies according to the fluid composition, service location, geometry, temperature, and so forth. In all cases of corrosion, the electrolyte must be present for the reaction to occur. In the oil and gas production industries, the major forms of corrosion include [17,18] sweet corrosion, sour corrosion, oxygen corrosion, galvanic corrosion, crevice corrosion, erosion corrosion, microbiologically induced corrosion, and stress corrosion cracking.

Sweet Corrosion (CO_2 Corrosion)

CO_2 corrosion has been a recognized problem in oil and gas production and transportation facilities for many years [19]. CO_2 is one of the main corroding agents in the oil and gas production systems [20]. Dry CO_2 gas is not itself corrosive at the temperatures encountered within oil and gas production systems but is so when dissolved in an aqueous phase through which it can promote an electrochemical reaction between steel and the contacting aqueous phase [21]. CO_2 will mix with the water, forming carbonic acid making the fluid acidic. CO_2 corrosion is influenced by temperature, increase in pH value, composition of the aqueous stream, presence of non-aqueous phases, flow condition, and metal characteristics [20,22] and is by far the most prevalent form of attack encountered in oil and gas production [2]. At elevated temperatures, iron carbide scale is formed on the oil and gas pipe as a protective scale, and the metal starts to corrode under these conditions. CO_2 corrosion can appear in two principal forms: pitting (localized attack that results in rapid penetration and removal of metal at a small discrete area) [23] and mesa attack (a form of localized CO_2 corrosion under medium-flow conditions) [24]. Figures 2 and 3 represent pitting corrosion and mesa attack, respectively.

Figure 2: Pitting corrosion.

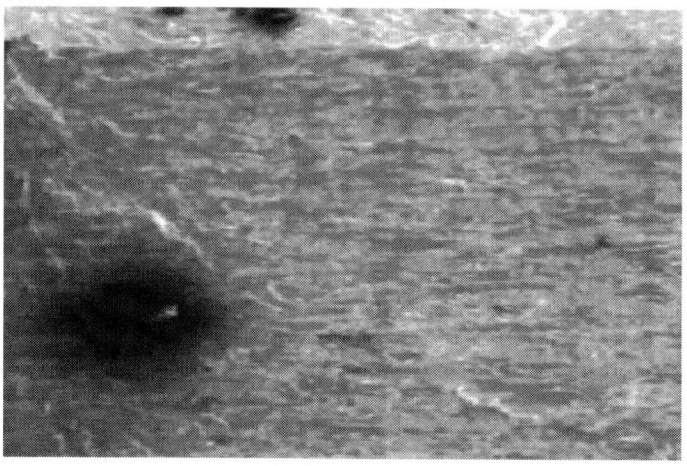

Figure 3: Mesa attack.

Various mechanisms have been postulated for the CO_2 corrosion process but all involve either carbonic acid or the bicarbonate ion formed on dissolution of CO_2 in water. The best known mechanism was postulated by de Waard et al. [25] and was given as

$$H_2CO_3 + e^- \rightarrow H + HCO_3^-$$
(8)

$$2H \rightarrow H_2$$
(9)

With the steel reacting

$$Fe \rightarrow Fe^{2+} + 2e^-$$
(10)

And overall

$$CO_2 + H_2O + Fe \rightarrow FeCO_3 \text{ (Iron carbonate)} + H2:$$
(11)

Sour Corrosion (H₂S Corrosion)

The deterioration of metal due to contact with hydrogen sulfide (H_2S) and moisture is called sour corrosion which is the most damaging to drill pipe. Although H_2S is not corrosive by itself, it becomes a severely corrosive agent in the presence of water [26], leading to pipeline embrittlement [20]. Hydrogen sulfide when dissolved in water is a weak acid, and therefore, it is a source of hydrogen ions and is corrosive. The corrosion products are iron sulfides (FeS_x) and hydrogen. Iron sulfide forms a scale that at low temperature can act as a barrier to slow corrosion [18]. The forms of sour corrosion are uniform, pitting, and stepwise cracking. Figure 4 is the diagram of an oil and gas pipeline under sour corrosion. The general equation of sour corrosion can be expressed as follows [27]:

$$H_2S + Fe + H_2O \rightarrow FeS_x + 2H + H_2O. \tag{12}$$

Figure 4: Oil and gas pipeline under sour corrosion.

Another probable mechanism for iron dissolution in aqueous solutions containing H_2S based on the formation of a mackinawite film, as proposed by Sun et al. [28], is shown in Figure 5.

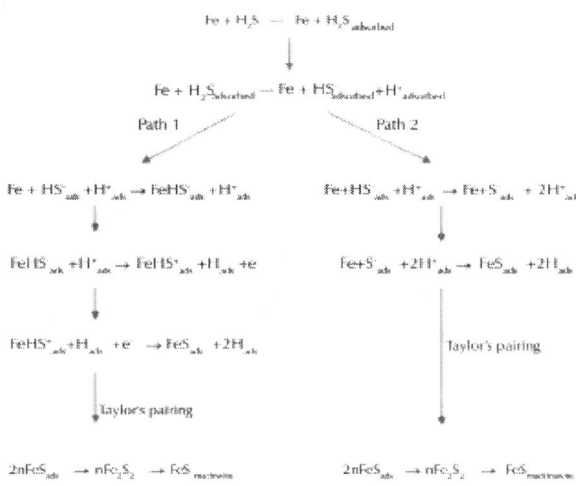

Figure 5: Probable mechanism for iron dissolution in aqueous solutions containing H2S.

Oxygen Corrosion

Oxygen is a strong oxidant and reacts with the metal very quickly. Oxygen dissolved in drilling fluids is a major cause of drill pipe corrosion. Oxygen ingress takes place in the well fluids through leaking pump seals, casing, and process vents and open hatches. As a depolarizer and electron acceptor in cathodic reactions, oxygen accelerates the anodic destruction of metal [29]. The high-velocity flow of drilling fluids over the surfaces of a drill pipe continues to supply oxygen to the metal and is destructive at concentrations as low as 5 ppb [30]. The presence of oxygen magnifies the corrosive effects of the acid gases (H_2S and CO_2). The inhibition of corrosion promoted by oxygen is difficult to achieve and is not practical in the drilling fluid system. The forms of corrosion associated with oxygen are mainly uniform corrosion and pitting-type corrosion. Figure 6 shows the diagrammatic representation of oxygen corrosion.

Figure 6: Oxygen corrosion.

Galvanic Corrosion

This type of corrosion occurs when two metallic materials with different nobilities (electrochemical potential) are in contact and are exposed to an electrolytic environment. In such situation, the metal with less or the most negative potential becomes the anode and starts corroding [20,31]. The anode loses metal ions to balance electron flow. Because metals are made up of crystals, many of such cells are set up, causing intergranular corrosion. Problems are most acute when the ratio of the cathode-to-anode area is large [18]. Figure 7 is the diagrammatic representation of the galvanic corrosion process, while Figure 8 is the galvanic corrosion resulting from placing a bronze sea strainer on an aluminum hose barb as part of the equipment used during oil and gas production.

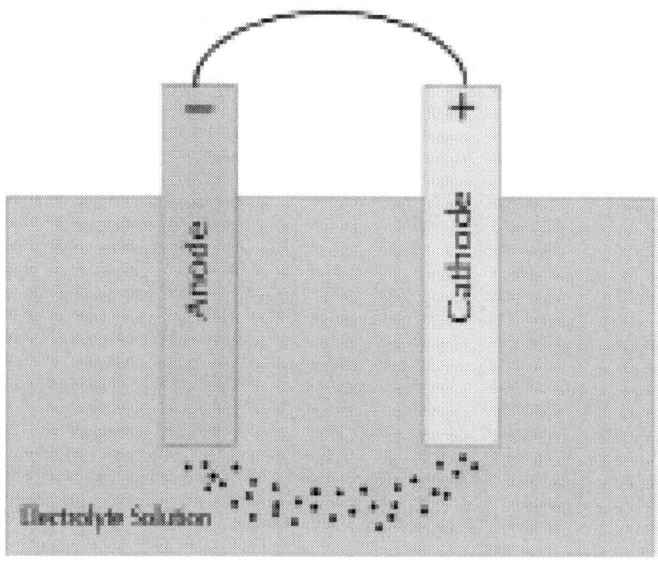

Figure 7: Galvanic corrosion process.

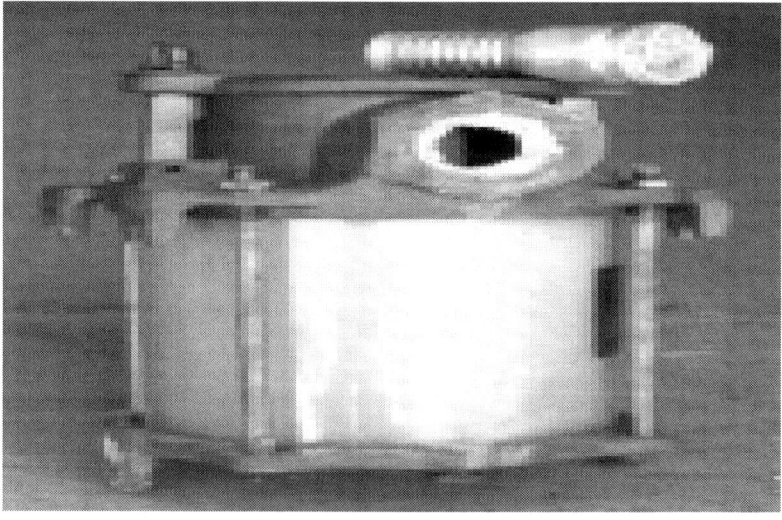

Figure 8: Galvanic corrosion resulting from placing a bronze sea strainer on an aluminum hose barb.

Crevice Corrosion

Crevice corrosion is normally a localized corrosion taking place in the narrow clearances or crevices in the metal and the fluid getting stagnant in the gap. This is caused by concentration differences of corrodents over a metal surface [1]. Electrochemical potential differences result in selective crevice or pitting corrosion attack. Oxygen dissolved in drilling fluid promotes crevice and pitting attack of metal in the shielded areas of drill string and is the common cause of washouts and destruction under rubber pipe protectors [32]. Figure 9 depicts an oil and gas pipeline under crevice corrosion.

Figure 9: Oil and gas pipeline under crevice corrosion.

Erosion Corrosion

The erosion corrosion mechanism increases corrosion reaction rate by continuously removing the passive layer of corrosion products from the wall of the pipe. The passive layer is a thin film of corrosion product that actually serves to stabilize the corrosion reaction and slow it down. As a result of the turbulence and high shear stress in the line, this passive layer can be removed, causing the corrosion rate to increase

[33]. The erosion corrosion is always experienced where there is high turbulence flow regime with significantly higher rate of corrosion [34] and is dependent on fluid flow rate and the density and morphology of solids present in the fluid [20]. High velocities and presence of abrasive suspended material and the corrodents in drilling and produced fluids contribute to this destructive process. This form of corrosion is often overlooked or recognized as being caused by wear [35].

Microbiologically Induced Corrosion

This type of corrosion is caused by bacterial activities. The bacteria produce waste products like CO_2, H_2S, and organic acids that corrode the pipes by increasing the toxicity of the flowing fluid in the pipeline [36]. The microbes tend to form colonies in a hospitable environment and allow enhanced corrosion under the colony. The formation of these colonies is promoted by neutral water especially when stagnant [20]. Numerous reports of the presence of microbes in reservoirs had been published [37-39]. Lazar et al. [38] found abundant microbial flora indigenous in oil field formation waters, which included species of *Bacillus, Pseudomonas, Micrococcus, Mycobacterium,Clostridium,* and *Escherichia. Escherichia* is reported to contain hydrogenase, an enzyme that utilizes molecular hydrogen and may be associated with cathodic hydrogen depolarization, causing corrosion of steel casings and pipes in the oil field [40]. Bacteria that form slime (some form of polysaccharides), such as *Achromobacter* sp., *Flavobacterium* sp., and *Desulfuricans* sp., will adhere to each other, forming a large mass. They also adhere to the walls of the pores, causing severe plugging problems at injection wells [39]. Microbiologically induced corrosion (MIC) is recognized by the appearance of a black slimy waste material or nodules on the pipe surface as well as pitting of the pipe wall underneath these deposits. Figures 10 and 11 represent the scanning electron microscopy (SEM) photograph of *Desulfovibrio desulfuricans* and a pipeline affected by MIC corrosion, respectively [41].

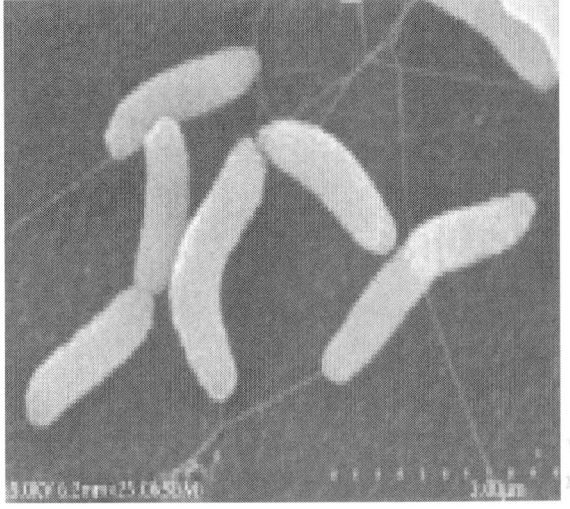

Figure 10: SEM photograph of *D. desulfuricans*.

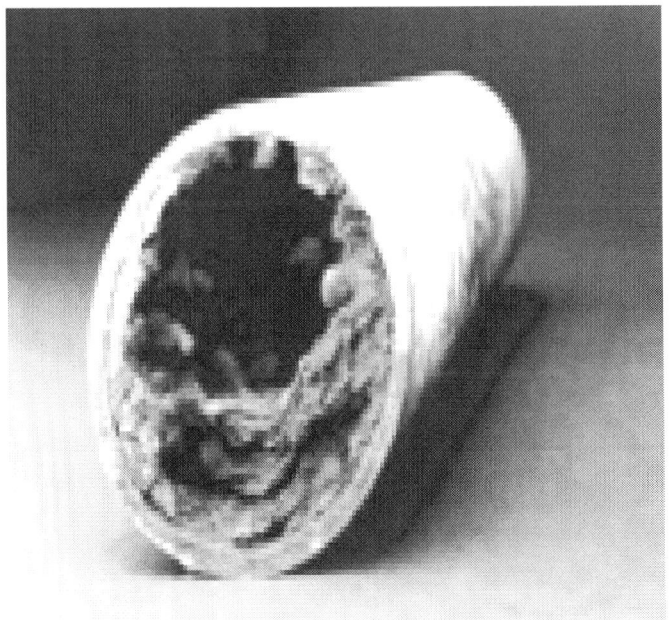

Figure 11: Pipeline affected by MIC corrosion.

Stress Corrosion Cracking

Stress corrosion cracking (SCC) is a form of localized corrosion which produces cracks in metals by simultaneous action of a corrodent and tensile stress. It propagates over a range of velocities from 10^{-3} to 10 mm/h depending upon the combination of alloy and environment involved. SCC is the cracking induced from the combined influence of tensile stress and a corrosive medium. The impact of SCC on a material seems to fall between dry cracking and the fatigue threshold of that material [42]. SCC in pipeline is a type of environmentally associated cracking. This is because the crack is caused by various factors combined with the environment surrounding the pipe. The most obvious identifying characteristic of SCC in a pipeline is high pH of the surrounding environment, appearance of patches, or colonies of parallel cracks on the external of the pipe [43]. Figure 12shows an oil and gas pipeline after being attacked by stress corrosion cracking.

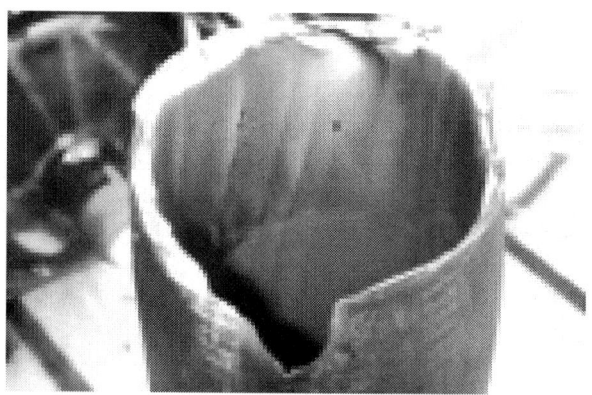

Figure 12: Oil and gas pipeline after being attacked by stress corrosion cracking.

Corrosion Mitigation in the Oil and Gas Industry

Oil field corrosion challenges are not static phenomena. Fluid characteristics change over time, resulting in systems becoming less responsive to established corrosion mitigation programs [3]. Within the sphere of corrosion control and prevention in the oil and gas industry, there are technical options such as cathodic and anodic protection, material selection, chemical dosing, and the application of internal and external coatings. It is widely recognized within the oil and gas industry that effective management of corrosion will contribute towards the maintenance of asset integrity and achieve optimization of mitigation, monitoring, and inspection costs [44]. While many methods have been advised to arrest these events, these methods can be broadly classed as follows [20]:

- Selection of appropriate materials
- Use of inhibitors
- Use of protective coatings
- Adequate corrosion monitoring and inspection
- Cathodic protection technique

When it is observed that the existing materials of construction are prone to corrosive attack, it is normally decided to change the materials of construction and select alternate materials to suit the specific need [20]. Stainless steels cover a wide range of alloys, each with a particular combination of corrosion resistance and mechanical properties. In oil and gas applications, many of these stainless steel grades are used, depending on the demands of the particular service environment. Applicable corrosion-resistant alloys in the oil and gas industry proposed by Smith [45] include 13Cr, Super 13Cr, 22Cr duplex, 25Cr duplex, 28Cr stainless steel, 825 nickel alloy, 625 nickel alloy, 2550 nickel alloy, and C276 nickel alloy. Johansson et al. [46] proposed a specialty stainless steel for solving corrosion problems in the oil and gas industry. The three stainless steels were LDX 2101, 254 SMO and 654 SMO (Outokumpu Stainless Steel and Alloys Company, Bergsnasgatan 11, 774 22 Avesta, Sweeden, +46 226 820 01). The resistance to localized corrosion of the stainless steels was estimated from the composition using the pitting resistance equivalent (PRE):

$$PRE = (\%Cr) + (3.3 \times \%Mo) + (16 \times \%N).$$

(13)

The chemical composition, mechanical properties, and results of various tests conducted for the recommended stainless steels for use in the oil and gas industries are summarized in Table 1[25].

Table 1: Chemical composition and mechanical properties of recommended stainless steels

Grade	Typical chemical composition (wt. %)						PRE	Microstructure	Rp0.2	Rm	A3	CPT ASTM G150	CPT ASTM G48 F	CCT ASTM G48 F
	Cr	Ni	Mo	C	N	Other			(MPa)	(MPa)	(%)			
LDX 2101	21.5	1.5	0.3	0.03	0.22	5 Mn	26	Duplex	450	650	30	17	15	<0
254 SMO	20	18	6.1	0.01	0.20	Cu	43	Austentic	300	650	40	87	65	35
654 SMO	24	22	7.3	0.01	0.50	Mn, Cu	56	Austentic	430	750	40	>90	>bp	60

ASTM American Society for Testing Materials, bp boiling point, CPT critical pitting temperature.

Popoola et al.

Popoola et al. International Journal of Industrial Chemistry 2013 4:35, doi:10.1186/2228-5547-4-35

The result of the drop evaporation test (DET) and sulfide stress cracking (SSC) testing of the stainless steels recommended by Johansson et al. [46] in NACE solution (Outokumpu Stainless Steel and Alloys Company, Bergsnasgatan 11, 774 22 Avesta, Sweeden, +46 226 820 01) (5% NaCl, pH 3, 1 bar p_{H2S}) for 720 h is presented in Table 2 [46].

Table 2: Drop evaporation and sulfide stress cracking tests for recommended stainless steels in oil and gas industry

Grade	Drop evaporation test				SCC testing			
	Wick test	40% CaCl2, 100°C	25% NaCl, bp	DET (%)	Cold work (%)	Stress (% of YS)	Temperature (°C)	Result
LDX 2101	No cracks	No cracks	No cracks	-	-	-	RT	-
254 SMO	No cracks	-	-	80	40 to 80	90	25	No cracks
654 SMO	-	-	-	100	0 to 80	100	25	No cracks

bp boiling point, RT room temperature, YS yield strength.

Popoola et al.

Popoola et al. International Journal of Industrial Chemistry 2013 4:35, doi:10.1186/2228-5547-4-35

Nalli [20] presented some of the commonly used materials (shown in Table 3) in the hydrocarbon and oil and gas industries based on a detailed study of process and operating conditions. He stated that a detailed study of flow conditions, corrosion mechanisms involved, and the expected life of a material is important before selecting a specific metal for the application. Mannan et al. [47] in their paper developed a new high-strength corrosion-resistant alloy 945 for oil and gas applications whose nominal composition was Fe-47Ni-20.5Cr-3Mo-2Cu-3Nb-1.5Ti. The alloy was developed to provide 125 ksi minimum yield strength and an excellent combination of ductility and impact strength. Craig [48] presented some alloys (shown in Table 4) whose applications in the oil and gas industry are majorly in the absence of oxygen.

Table 3: Recommended materials in the oil and gas industry

Material specification	Oil and gas applications
Carbon steels	Bulk fluids, crude pipelines, flow lines, water and steam injection lines, production and test separators, KO drums, storage tanks
Low- and medium-alloy steels	Well head items, chokes, manifolds and well components with sour and high-temperature applications
Straight chromium steels (chromium 12% to 18%)	Christmas trees, well heads, downhole rods, valves and casing pipes
Chromium-nickel steels (chromium >18%, nickel >8%)	Valve trims, instruments and materials of separators and tanks, low-chloride levels
Nickel steels (2.5%, 3.5%, 9% nickel)	Rarely used in oil and gas sectors, LNG storage tanks, piping and pumps
Duplex stainless steels (22% chromium duplex, 25% chromium super, duplex)	Piping, vessel and tank internals where a very high level of chlorides is present
Nickel-chrome (inconels) Ni-Cr-Fe alloys	Well head and flow lines, manifolds with high sour and temperature applications
Nickel-iron (incolys) Ni-Fe-Cr alloys	Well head and flow lines, manifolds with high sour and temperature applications

KO knock out, LNG liquefied natural gas.

Popoola et al.

Popoola et al. International Journal of Industrial Chemistry 2013 4:35, doi:10.1186/2228-5547-4-35

Table 4: Chemical composition of recommended materials in the oil and gas industry

Alloys	Nominal composition							Oil and gas application
	Cr	Ni	Mo	Fe	Mn	C	N	
13 Cr	13	-	-	Balanced	0.8	0.2	-	Corrosion resistance in CO2/NaCl environments in the absence of O2 and H2S
316	17	12	2.5	Balanced	1.0	0.04	-	Frequently used for oil field applications in the complete absence of oxygen
22 Cr	22	5	3	Balanced	1.0	0.1	0.1	Susceptible to localized corrosion in the presence of small amounts of O2 and H2S
25 Cr	25	7	4	Balanced	1.0	0.1	0.3	Corrosion resistance in H2S/CO2 environments in the absence of elemental sulfur

Popoola et al.

Popoola et al. International Journal of Industrial Chemistry 2013 4:35, doi:10.1186/2228-5547-4-35

Use of Inhibitors

Inhibitors are chemicals that are used to protect the surface of metals used in oil and gas industries to prevent corrosion. They protect the surface of metals either by merging with them or by reacting with the impurities in the environment that may cause pollution [49]. A corrosion inhibitor may act in a number of ways: It may restrict the rate of the anodic process or the cathodic process by simply blocking active sites on the metal surface. Alternatively, it may act by increasing the potential of the metal surface so that the metal enters the passivation region where a natural oxide film forms. A further mode of action of some inhibitors is that the inhibiting compound contributes to the formation of a thin layer on the surface which stifles the corrosion process [50].

Factors to be considered before using a corrosion inhibitor in the oil and gas industry include toxicity, environmental friendliness, availability, and cost. Organic corrosion inhibitors are more effective than inorganic compounds for protection of steels in acid media. A review of literature on high-temperature acid corrosion inhibitors revealed that the effective corrosion inhibitors for oil well acidization include acetylene alcohols, quaternary ammonium salts, aldehydes, amines, etc.[49]. Table 5 shows a list of recommended inhibitors by previous researchers and their places of applicability in the oil and gas industries.

Table 5: Recommended inhibitors for oil and gas applications

Inhibitors	Oil and gas applications
3-Phenyl-2-propyn-1-ol	API J55 oil field tubing in HCl solutions over a wide range of conditions [51]
Hydrazides and thiosemicarbazides of fatty acids with 11, 12, and 18 carbon atoms	Mild steel and oil well steel (N80) in boiling 15% hydrochloric acid solution [52]
Mixture of ketones, quinolinium salts, and formic acid	Oil field tubular goods to temperatures as high as 400°F (204°C) in hydrochloric [53]
2-Undecane-5-mercapto-1-oxa-3,4-diazole	Mild steel in 15% HCl at 105 ± 2°C and N80 steel in 15% HCl containing 5,000 ppm of 2-undecane-5-mercapto-1-oxa-3,4-diazole [54]
2-Heptadecene-5-mercapto-1-oxa-3,4-diazole	
2-Decene-5-mercapto-1-oxa-3,4-diazole	

Dibenzylidene acetone	N80 steel and mild steel in HCl [55]
Di-N-dimethylaminobenzylidene acetone	
Methoxy phenol and nonyl phenol	N80 steel in 15% HCl at different exposure periods (6 to 24 h) and temperatures (30°C to 110°C) [56]
N-(5,6-diphenyl-4,5-dihydro-[1,2,4]triazin-3-yl)-guanidine	Mild steel in 1 M hydrochloric acid and 0.5 M sulfuric acid [57]
6-Benzylaminopurine	Cold rolled steel in 1.0 to 7.0 M H2SO4 at 25°C to 50°C [58]
Mixture of synthetic magnetite and ferrous gluconate	Oil well steel (N80) in 50 mg/l sulfide concentration at various pH (5.5 to 11.5) and at high-temperature pressure conditions [59]
Rosin amide imidazoline	N80 and P110 carbon steels in CO2-saturated simulated formation water [60]

Popoola et al.

Popoola et al. International Journal of Industrial Chemistry 2013 4:35, doi:10.1186/2228-5547-4-35

Miksic et al. [61] had evaluated several types of corrosion inhibitors for the petroleum industry under various flow conditions. Active ingredients of the inhibitors included long-chain amines, fatty amides, imidazolines, fatty acids, and their salts. Inhibitors were tested at the concentration range of 50 to 200 ppm in the electrolyte and electrolyte/hydrocarbon mixture in the presence of CO_2 and H_2S in static and dynamic conditions. These products provide a very high level of protection for steel subjected to a broad range of corrosive attack and flow restriction from moisture, condensation, oxygen, carbon dioxide, hydrogen sulfide, and other corrosive contaminants. Unlike conventional methods, such as filming amine-based corrosion inhibitors, an injection of a volatile corrosion inhibitor (VpCl)-based material into any part of the system will set the VpCl to work immediately with a self-replenishing mono-molecular protective layer [62].

VpCl technology is an environmentally safe and cost-effective option for corrosion protection. VpCls form a physical bond on the metal surface and create a barrier layer to protect against aggressive ions. The barrier reheals and self-replenishes, and can be combined with other functional properties for added protective capabilities. It can be used in pipelines, oil and gas wells, refinery units, and fuels. In addition, these VpCI-based anti-corrosion additives have been designed to work well in multiphase flow systems in conjunction with different drag reducers. These different combinations of corrosion inhibitors and drag reducers provide systems with improved water flow and corrosion protection of pipelines carrying water or the mixture of hydrocarbon and water. All of these will lead to energy saving in oil production and increased overall recoverable reserves. The reduction of operating pressure will in turn give a lower back pressure in the well head and lead to additional oil production, enabling a substantial annual revenue increase. The pie chart showing the world consumption of corrosion inhibitors on a value basis is given in Figure 13 [63].

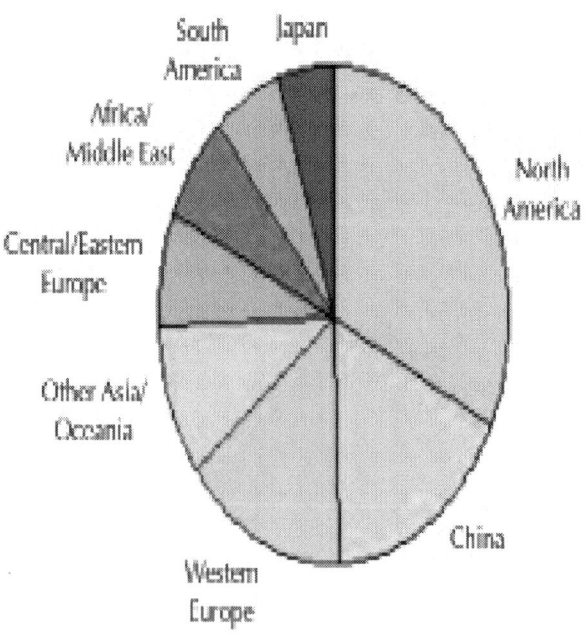

Figure 13: Pie chart showing the world consumption of corrosion inhibitors.

Use of Protective Coatings

A protective layer or barrier on the material to avoid the direct contact with the process media will enhance the material and equipment life. The barrier layer can be paint, a coating or a lining, or a metallic lining or metallic sheets. There are also non-metallic linings like fiber glass, glass flake, epoxy, and rubber which are normally carried out on the equipment like separators, knock-out drums, and storage tanks. Nickel, zinc, and cadmium coatings are also preferred at times on certain components like flanges and bolting [20]. The Phillips Ekofisk wells with low levels of H_2S, 90 lb in.$^{-2}$ CO_2, and up to 30,000 ppm chloride levels were completed with N-80 tubing. Even with batch inhibition, the tubing lasted only 19 months before it became perforated, and therefore, an extensive coating program was undertaken [64]. Where no inhibitor was injected, the coated tubing still only lasted about 19 months. Plastic coating on N-80 pipe with inhibitor batch treatment every 30 days gave a tubing life of 7 years [65].

Fusion-bonded epoxy (FBE) and a three-layer polyolefin (3LPO) (polyethylene or polypropylene (PP)) are currently the most widely used external anti-corrosion coating systems. Figures 14 and15 represent 3LPO and FBE coatings, respectively [66]. Single-layer FBE has been more popular in North America, Saudi Arabia, and the UK; dual-layer FBE is in favor in Australia; and 3LPO coatings dominate the rest of the world's pipe coating market [67]. Bredero Shaw, a world leader in pipe coating solutions and with more than 75 years of experience, over 27 pipe coating facilities on six continents, and the largest team of technical and service specialists in the business, presented several unique advanced and proven pipeline coating technologies and services designed to protect pipelines for onshore and offshore applications. These include [68]

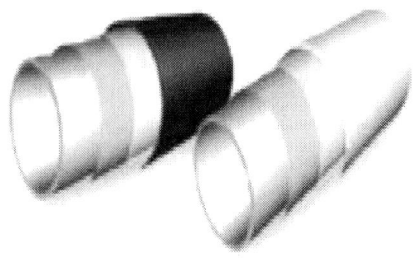

Figure 14: 3LPO coating.

- High Performance Composite Coating system (HPCC)
- Low Temperature Application Technology for Powder Coating on High Strength Steel
- Thermotite Flow Assurance Coating Technology

Figure 15: FBE coating.

High Performance Composite Coating System

The HPCC is a single-layer, all-powder-coated, multicomponent coating system consisting of a FBE base coat, a medium-density polyethylene outer coat, and a tie layer containing a chemically modified polyethylene adhesive. All materials of the three components of the composite coating are applied using an electrostatic powder coating process. The tie layer is a blend of adhesive and FBE with a gradation of FBE concentration. Thus, there is no sharp and well-defined interface between the tie layer and either of the FBE base coat or the polyethylene outer coat. Figure 16shows a cross section of the composite coating with a standard total thickness of 750 µm (30 mil).

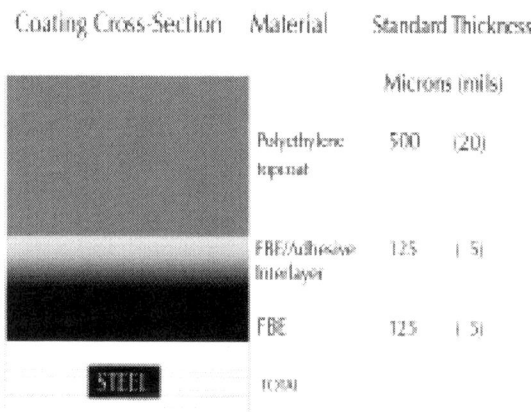

Figure 16: Cross section of the composite coating.

Low Temperature Application Technology for Powder Coating on High Strength Steel

High yield strength steel is often used for constructing oil and steel pipes because it allows the reduction of pipe wall thickness. This technology is applied in frontier areas like the Mackenzie-Beaufort area, the Arctic Islands, and the Labrador basins. Due to the incredibly cold temperatures in these areas, many issues concerning the strength and flexibility of pipelines have developed. Pipes that would be flexible enough to bend and shift whenever frost heaving occurred are necessary to sustain consistent flow. Frost heaving occurs whenever the ground changes from hard during the winter months to soft during warmer temperatures. This has the effect of causing any material that is built underground to shift with it. To address this issue, more flexible and high yield strength grades of steel such as X80 or higher were developed. A coating on the high-strength steel pipes for Frontier areas should withstand the extremely cold temperatures and retain the flexibility needed to protect the pipes. In addition, there is a need for a coating with indentation and impact resistance at $-40°C/-50°C$. This coating method had been used in frontier oil and gas in Canada.

Thermotite Flow Assurance Coating Technology

The Thermotite technology consists of a multilayer polypropylene composite FBE as the layer to the steel. Specific requirements for protection or thermal insulation are taken care of through the bespoke system design. Resistance to the effects of compression and creep, typical for deep water and high temperature, can be catered for by adjusting the density and nature of the layers. Figure 17 shows the Thermotite five-layer system build-up (Bredero Shaw Company, 25 Bethridge Road, Toronto, Ontario, Canada M9W 1M7) [68]. The three-layer anti-corrosion coating is applied by a side or cross-head extrusion process and the quality tested and approved, prior to the application of the thermal insulation layers (two-layers; PP foam and outer shield). The thermal layers and outer shield or weight coating polypropylene, are applied simultaneously in the thermal insulation lines, by a cross-head extrusion process. The method secures a fixed outer diameter and homogenous foam structure with no air inclusions.

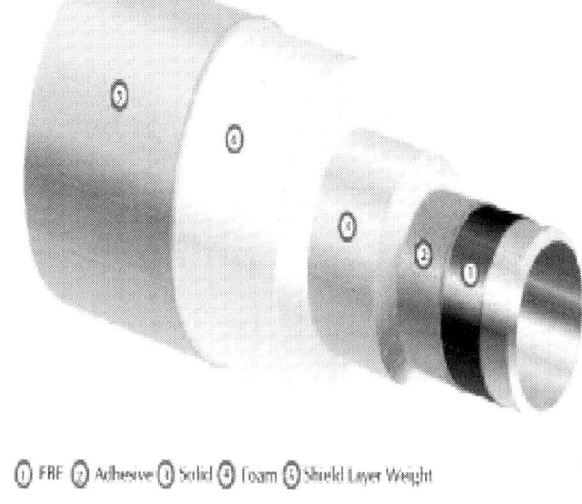

① FRF ② Adhesive ③ Solid ④ Foam ⑤ Shield Layer Weight

Figure 17: Thermotite five-layer system buildup.

Cathodic Protection Technique

The first application of cathodic protection dates back to 1824, long before its theoretical foundation was established, and is credited to Sir Humphrey Davy [69]. Cathodic protection is a method to reduce corrosion by minimizing the difference in potential between anode and cathode. This is achieved by applying a current to the structure to be protected (such as a pipeline) from some outside source. When enough current is applied, the whole structure will be at one potential; thus, anode and cathode sites will not exist [70]. It is normally used in conjunction with coatings and can be considered as a secondary corrosion control technique. The cathodic protection system can be designed to prevent both oxygen-controlled and microbiologically controlled corrosion [71]. The two methods of applying cathodic protection include [72]

• Sacrificial (or galvanic) anode cathodic protection (SACP)
• Impressed current cathodic protection (ICCP)

The main difference between the two is that ICCP uses an external power source with inert anodes and SACP uses the naturally occurring electrochemical potential difference between different metallic elements to provide protection.

Sacrificial Anode Cathodic Protection

In this type of application, the naturally occurring electrochemical potentials of different metals are used to provide protection. Sacrificial anodes are coupled to the structure under protection and conventional current flows from the anode to the structure as long as the anode is more active than the structure. As the current flows, all the corrosions occur on the anode which sacrifices itself in order to offer protection from corrosion to the structure. Figure 18 is the diagram representing sacrificial anode cathodic protection [1].

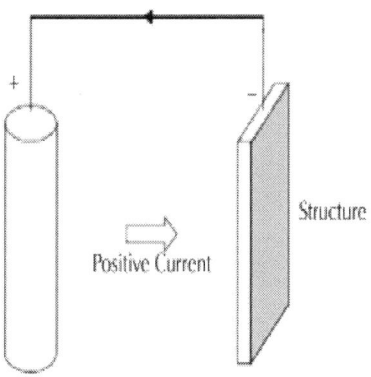

Figure 18: Sacrificial anode cathodic protection.

Impressed Current Cathodic Protection (ICCP)

In impressed current cathodic protection, the current is impressed or forced by a power supply. The power source must be able to deliver direct current, and examples are transformer rectifier units, solar generating units, or thermoelectric generators. The anodes are either inert or have low consumption rates and can be surrounded by carbonaceous backfill to increase efficiency and decrease costs. Typical anodes are titanium coated with mixed metal oxide or platinum, silicon iron, graphite, and magnetite. Laoun et al. [73] had applied impressed current cathodic protection to a buried pipeline by solar energy using photovoltaic generator as the power source. Table 6shows the characteristics of the buried pipeline [74]. They concluded that the method is applicable for various types of grounds and that the output current is high enough to protect the pipeline with low costs. Figure 19 represents the diagram of the ICCP used [73].

Table 6: Characteristics of the pipeline examined for ICCP

	Description
Material	Steel X60
Length	292 km
External diameter	0.762
Surface to protect	699,020 m2
Isolation resistance	800 Ω m
Linear isolation resistance	3340 Ω m
Longitudinal isolation resistance	7.49 × 10−6 Ω m
Attenuation coefficient	47.35 × 10−6 m−1
Characteristic resistance	158.16 × 10−3 Ω

Popoola et al.

Popoola et al. International Journal of Industrial Chemistry 2013 4:35, doi:10.1186/2228-5547-4-35

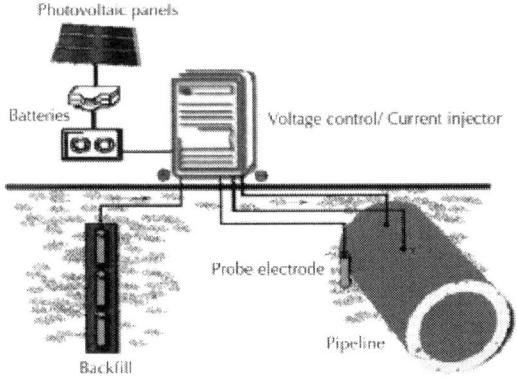

Photovoltaic panels

Batteries

Voltage control/ Current injector

Probe electrode

Pipeline

Backfill

Figure 19: Diagram of the ICCP used.

Adequate Corrosion Monitoring and Inspection

Corrosion monitoring is the practice of measuring the corrosivity of process stream conditions by the use of probes (mechanical, electrical, or electrochemical devices) which are inserted into the process stream and continuously exposed to the process stream condition. Corrosion monitoring techniques alone provide direct and online measurement of metal loss/corrosion rate in oil and process systems [75]. One of the methods is to carry out the on-stream inspection by doing the wall thickness measurements periodically on fixed and vulnerable locations on the equipment, piping, and pipelines to assess the material conditions and corrosion rates [75]. Also, corrosion is monitored by placing electronic probes in the pipelines and by measuring the change in the electric resistance in the probe coil. The cross-country pipelines are normally checked with intelligent pigging operations like magnetic flux or ultrasonic pigs. These pigs will detect the internal conditions of the pipeline and corrosion conditions on the pipe wall thickness and also indicate the wall thickness available on the pipe wall [20].

Most of the equipment like separators, drums, and heaters are checked for corrosion during annual shutdown and turnaround operations. Based on the physical assessment of the material conditions, corrective action is initiated to change the material or replace the equipment or at times do temporary repair work before replacement is carried out. In practice, it is observed that physical inspection is the best method of monitoring corrosion and assessing the material conditions. Other areas where corrosion monitoring and inspection are necessary in the oil and gas industry include drilling mud systems, digesters, water wash systems, flow lines, transport pipelines, desalters, sour water strippers, crude overheads, and many more [3]. Figure 20 is the framework for successful corrosion management [44].

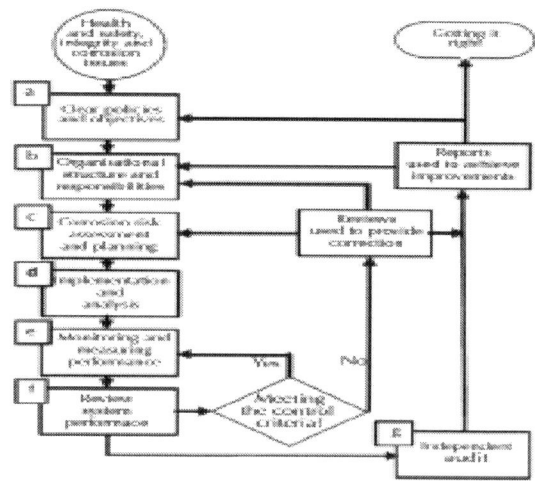

Figure 20: Framework for successful corrosion management.

DISCUSSION

From corrosion mitigation in the oil and gas industry earlier presented, solutions had been provided for various corrosion types discussed in the "Corrosion types and associated agents in the oil and gas industry" section of this paper. The critical pitting temperature using American Society for Testing Materials (ASTM) G150 and ASTM G48 method E confirmed LDX 2101 to be resistant to pitting corrosion (a form of CO_2 corrosion) in the oil and gas industry. When stresses are present in a chloride-containing environment, steels are prone to chloride-induced stress corrosion cracking (SCC). The results of a large number of tests demonstrated the superior resistance of LDX 2101 to SCC compared to standard austenitic grades in all these types of tests[46]. Also, the superaustenitic grade stainless steels (254 SMO and 654 SMO) presented in Table 1also showed excellent resistance to SCC and SSC. NACE allowed the use of both solution-annealed and cold-worked 254 SMO and 654 SMO at any temperature up to 171°C in sour environments, if the partial pressure of hydrogen sulfide does not exceed 15 psi (1 bar), the chloride content does not exceed 5,000 ppm, and the hardness is not greater than HRB 95 for the solution-annealed material and HRC 35 for the cold-worked material [76]. Nevertheless, 254 SMO had been shown to be susceptible to crevice corrosion at temperatures above 30°C in chlorinated seawater, while the extreme corrosion resistance of 654 SMO makes it an alternative to titanium and nickel-based alloys in many environments where crevice corrosion is possible. Both groups of stainless steels find extensive use in seawater applications, demanding process fluids and sour service. Thus, both 254 SMO and 654 SMO have an important place in the material solutions for the oil and gas industry as they tackle the problem of sour corrosion caused by a H_2S corroding agent.

The materials presented by Nalli [20] in Table 3 have various applications in different equipments of the oil and gas industry. The most prominent of the materials recommended are inconels and incolys which are Ni-Cr-Fe alloys. Though their nominal compositions were not presented, they are found to be very applicable in well head and flow lines with high sour and temperature applications. The materials (13 Cr, 316, 22 Cr, and 25 Cr) recommended by Craig [48] which are presented in Table 4 are also applicable to tackle corrosion in the oil and gas industry but are only active in the absence of oxygen, H_2S, and elemental sulfur. The alloy (13 Cr) is corrosion resistant in CO_2/NaCl environments in the absence of O_2 and H_2S, while 25 Cr is corrosion resistant in H_2S/CO_2 environments in the absence of elemental sulfur. The new alloy presented by Mannan et al. [47] with nominal composition Fe-47Ni-20.5Cr-3Mo-2Cu-3Nb-1.5Ti was developed by the method of homogenized vacuum induction melting. The mechanical properties and microstructure exhibited by the alloy after corrosion testing indicated that it could resist stress corrosion cracking, galvanically induced hydrogen stress cracking, and sulfide stress cracking [77].

Inhibitors had been shown to be one of the major tools for tackling corrosion in the oil and gas industries. They execute this task by protecting the surface of metals either by merging with them or by reacting with the impurities in the environment that may cause pollution [49]. Various inhibitors applicable in the oil and gas industries had been presented in Table 5. Majority of them (6-benzylaminopurine, rosin amide imidazoline, methoxy phenol, nonyl phenol, and so on) had been shown to be major tools in tackling oxygen corrosion [63] through the removal of oxygen from the fluid media and thus improves the chances of corrosion resistance of materials in contact[20]. The pie chart presented in Figure 11 shows that North America had the largest consumption of corrosion inhibitors as of 2008. This is reflected in the statement made by Tuttle [13] that corrosion costs US industries alone an estimated $170 billion a year in which the oil and gas industry takes more than half of these costs. This means that the USA had been looking for ways of cutting down expenses lost to corrosion since 1987. Corrosion management has improved over the past several decades; the USA must find more and better ways to encourage, support, and implement optimal corrosion control practices [78].

Corrosion control through the use of inhibitors is not recommended for hostile wells because this option has high operating cost implications over the full field life owing to the operating costs of inhibitor injection and the higher frequency of workovers. Nevertheless, there are concerns about the efficacy of inhibitors in controlling sulfide stress cracking in carbon steels [45]. However, the new impressed current cathodic protection method presented is a better method of tackling these problems as the process has a broad potential protection and the system is adaptable for different materials constituting the oil and gas pipelines. The method also has controllable output current that is high enough to protect pipelines with low costs. Both methods of the cathodic protection presented can be designed to prevent oxygen-controlled and microbiologically controlled corrosion.

The Low Temperature Application Technology for Powder Coating on High Strength Steel presented is very unique as the method is applicable in frontier areas such as the Mackenzie-Beaufort area, the Arctic Islands, and the Labrador basins due to the incredibly cold temperatures in these areas. Frontier oil and gas in Canada provides one example of this. Also, the Thermotite insulation and flow assurance coating systems have been used in many major offshore pipe applications in the North Sea, Gulf of Mexico, South China Sea, and other areas, representing a total insulated length of 660 km and a total coated length of 1,005 km [68]. Houghton et al. [64] and Hovsepian et al.[65] had shown that the use of inhibitors coupled with the use of protective coatings to avoid direct contact with process media are efficient means of mitigating corrosion in the oil and gas industry. The examined case study of Phillips Ekofisk wells with low levels of H_2S, 90 lb in.$^{-2}$ CO_2, and up to 30,000 ppm chloride levels showed that the N-80 pipe used gave a tubing life of 7 years (longer) for plastic coating with inhibitor batch treatment every 30 days, while it only lasted for

19 months before it became perforated when only batch inhibition was used.

However, the best way to check corrosion is by visual inspection and checking up the material degradation periodically. Undoubtedly, understanding the corrosion mechanism is very important before considering various material options for the applications. It should be clearly understood that no particular material is the cure for all the evils of corrosion. Each and every case has to be considered in its totality before a decision is made on the proper materials. The framework for successful corrosion management is presented in Figure 20, while various forms of corrosion monitoring techniques are presented elsewhere [75].

CONCLUSIONS

Corrosion is a stochastic, probabilistic phenomenon that requires interdisciplinary concepts that incorporate surface science, metallurgy/ materials science, electrochemistry, thermodynamics and kinetics, mechanics, hydrodynamics, and chemistry. It costs the oil and gas industry tens of billions of dollars in lost income and treatment costs every year. It should be noted that the damage caused by corrosion is not only in the oil and gas industry but also in other major areas like building construction, transportation, production and manufacturing, and so on. Thus, corrosion is a world problem which everybody must find a solution to as it covers many areas in our daily needs. In this paper, comprehensive review of corrosion in the oil and gas industry had been considered. Various corrosion types and their associated corroding agents in the oil and gas industry had been examined alongside with ways of mitigating them. However, the principles of corrosion must be understood in order to effectively select materials and to design, fabricate, and utilize metal structures for the optimum economic life of facilities and safety in oil and gas operations. Also, it should be clearly understood that no particular material is the cure for all the evils of corrosion. Each and every case has to be considered in its totality before a decision is made on the proper materials. Consultations with process, operation, material, and corrosion engineers are necessary in the fitness of things to save millions to fight the corrosion menace.

AUTHORS' CONTRIBUTIONS

LTP participated in the critical review of the paper. ASB provided information on the recommended inhibitors for oil and gas application with the cathodic protection technique. GKL gave diagrams for various corrosion mechanisms. LTP, BG, and ASG did the revision of the manuscript. All authors read and approved the final manuscript.

ACKNOWLEDGMENTS

The authors would like to acknowledge Professor (Emeritus) A. A. Susu of the University of Lagos, Nigeria for the path he has laid down for us in the field of chemical engineering in Nigeria and for his support and words of encouragement on the success of this paper. We also like to thank the reviewers of this paper (anonymous) under the International Journal of Industrial Chemistry for making this paper to be academically standard.

REFERENCES

1. Roberge PR (2000) Handbook of corrosion engineering New York: McGraw-Hill

2. Kermani MB, Smith LM (1997) CO_2 corrosion control in oil and gas production: design considerations London: The Institute of Materials, European Federation of Corrosion Publications

3. Champion Technologies (2012) Corrosion mitigation for complex environments Houston: Champion Technologies

4. Corbin D, Willson E (2007) New technology for real-time corrosion detection USA: Tri-service corrosion conference

5. Lusk D, Gupta M, Boinapally K, Cao Y (2008) Armoured against corrosion. Hydrocarb Eng. 13:115–118

6. Oyelami BO, Asere AA Mathematical modelling: An application to corrosion in a petroleum industry. NMC Proceedings Workshop on Environment Abuja, Nigeria: National Mathematical Centre

7. Miller D (1990) Corrosion control on aging aircraft: what is being done? Mater Perform 29:10–11

8.   Hoffman C (1997) 20,000-Hour tuneup. Air and Space 12:39–45

9.   Up Front (1992) The Cost of corrosion in the EEC. Materials performance. 31:3

10.   Trethewey KR, Roberge PR (1995) Corrosion management in the twenty-first century. British Corro J. 30:192–197

11.   Uhlig HH (1949) The cost of corrosion in the United States. Chem and Engng News. 27:2764

12.   Simons MR (2008) Report of offshore technology conference (OTC) presentation NACE International oil and gas production

13.   Tuttle RN (1987) Corrosion in oil and gas production. J of Petrol Technol 39:756–762

14.   http://www.touchbriefings.com/pdf/30/exp032_p_12Nyborg.pdf

15.   Nimmo B, Hinds G (2003) Beginners guide to corrosion Teddington: NPL

16.   Dean F, Powell S (2006) Hydrogen flux and high temperature acid corrosion NACExpo 2006 conference

17.   Oxford WF, Foss RE (1958) Corrosion of oil and gas well equipment Dallas: Division of Production, American Petroleum Institute

18.   Brondel D, Edwards R, Hayman A, Hill D, Mehta S, Semerad T (1994) Corrosion in the oil industry. Oilfield Rev

19.   Kermani MB, Harrop D (1996) The impact of corrosion on the oil and gas industry 11: SPE Production Facilities pp 186–190

20.   Nalli K (2010) Corrosion and its mitigation in the oil and gas industry PM-Pipeliner Report

21.   Dugstad A (1992) The importance of FeCO3 supersaturation on the CO2 corrosion of carbon steels, corrosion "92, paper 14 Houston: NACE

22.   Gatzky LK, Hausler RH (1984) A novel correlation of tubing corrosion rates and gas production rates. Adv in CO2 Corro 1:87

23.   Schmitt G (1984) Fundamental aspects of CO2 corrosion Houston: NACE p 10

24.   Dunlop A, Hassel HL, Rhodes PR (1984) Fundamental considerations in sweet gas well corrosion Houston: NACE p 52

25.   de Waard C, Lotz U (1994) Prediction of CO2 corrosion of carbon steel London: The Institute of Materials

26. Ray JD, Randall BV, Parker JC (1978) Use of reactive iron oxide to remove H2S from drilling fluid Houston: 53rd Annu. Fall Tech. Conf. of AIME

27. Chilingar GV, Beeson CM (1969) Surface operations in petroleum production New York: American Elsevier p 397

28. Sun W (2006) Kinetics of iron carbonate and iron sulfide scale formation in CO2/H2S corrosion. PhD dissertation

29. Weeter RF (1965) Desorption of oxygen from water using natural gas for counter-current stripping. J Petrol Technol 17(5):51

30. Snavely ES (1971) Chemical removal of oxygen from natural waters. J Petrol Technol 23(4):443–446

31. Martin RL (1982) Use of electrochemical methods to evaluate corrosion inhibitors under laboratory and field condition Manchester: UMIST conference of electrochemical techniques

32. Hudgins CM (1969) A review of corrosion problems in the petroleum industry. Mater Prot 8(1):41–47

33. Hassani S, Roberts KP, Shirazi SA, Shadley JR, Rybicki EF, Joia C (2012) Flow loop study of NaCl concentration effect on erosion corrosion of carbon steel in CO$_2$ saturated systems. CORRO 68:2

34. Sami AA, Mohammed AA (2008) Study synergy effect on erosion-corrosion in oil and gas pipelines. Engng and Technol 26:9

35. Bertness TA (1957) Reduction of failures caused by corrosion in pumping wells. API Dril Prod Pract 37:129–135

36. Ossai CI (2012) Advances in asset management techniques: an overview of corrosion mechanisms and mitigation strategies for oil and gas pipelines. ISRN Corro 570143

37. Crawford PB (1983) Donaldson EC, Clark JB (ed) Possible reservoir damage from MEOR,. Springfield, VA: Proceedings of 1982 international conference on microbial enhancement of oil recovery, Afton, OK, May 16–21. pp 76–79 NTIS

38. Lazar I, Constantinescu P (1985) Field trials results of microbial enhanced oil recovery. In: Zajic JE, Donaldson EC (ed) Microbes and oil recovery, El Paso: Bioresources Publications pp 122–143

39. Singer ME (1985) Microbial biosurfactants. In: Zajic JE, Donaldson EC (ed) Microbes and oil recovery, El Paso: Bioresources Publications pp 19–38

40. Gates GPL, Parent CF (1976) Water quality control presents challenge in giant Wilmington Field. Oil Gas J 74(33):115–126

41. http://www.lbl.gov/Publications/Currents/Archive/Apr-30-2004.html

42. Wilhelm SM, Kane RD (1987) Status report: corrosion resistant alloys. Petrol Engng Int 3641

43. Baker MJ (2004) Stress corrosion cracking study. http://www.polyguardproducts.com/products/pipeline/TechReference/SCCReport-FinalReportwithDatabase.pdf webcite

44. Energy Institute (2008) Guidance for corrosion management in oil and gas production and processing London: Annual Report

45. Smith L (1999) Control of corrosion in oil and gas production tubing. British Corro J. 34(4):247

46. Johansson E, Pettersson R, Alfonsson E, Weisang-Hoinard F (2010) Specialty stainless for solving corrosion problems in the oil and gas industry. Offshore World 40:

47. Mannan S, Patel S (2008) A new high strength corrosion resistant alloy for oil and gas applications New Orleans: Paper presented at NACE Corrosion

48. Craig BD (1995) Selection guidelines for corrosion resistant alloys in the oil and gas industry. NiDI Tech Series 10073:1–8

49. Rajeev P, Surendranathan AO, Murthy CSN (2012) Corrosion mitigation of the oil well steels using organic inhibitors – a review. J Mater Environ Sci 3(5):856–869

50. Graeme W (2010) Corrosion protection of metals in marine environment J. Metal Corrosion Protection, Chemistry Department, University of Auckland

51. Growcock FB, Lopp UR (1988) The inhibition of steel corrosion in hydrochloric acid with 3-phenyl-2-propyn-1-ol. Corro Sci. 28:397–410

52. Quraishi MA, Jamal D (2000) Fatty acid triazoles novel corrosion inhibitors for oil well steel (N-80) and mild steel. JAOCS 77:1107–1112

53. Frenier WW, Schlumberger D (1989) Acidizing fluids used to stimulate high temperature wells can be inhibited using organic

chemicals Houston: Conference paper, SPE international symposium on oil field chemistry

54. Quraishi MA, Jamal D (2001) Corrosion inhibition of fatty acid oxadiazoles for oil well steel (N-80) and mild steel. Materials Chem and Phy. 71:202–205

55. Athar A, Sli MN, Quraishi MA (2001) A study of some new organic inhibitors on corrosion of N-80 and mild steel in hydrochloric acid. Anti-Corro Meth and Mat. 48:251–255

56. Vishwanatham S, Haldar N (2007) Corrosion inhibition of N-80 steel in hydrochloric acid by phenol derivatives. Indian J of Chem Technol 14:501–506

57. Khaled KF (2008) New synthesized guanidine derivative as a green corrosion inhibitor for mild steel in acidic solutions. Int J Electrochem Sci 3:462–475

58. Xianghong L, Deng S, Fu H, Guannan MU (2009) Inhibition effect of 6-benzylaminopurine on the corrosion of cold rolled steel in H_2SO_4 solution. Corro Sci. 51:620–634

59. Amosa MK, Mohammed IA, Yaro SA, Arinkoola O, Ogunleye OO (2010) Corrosion inhibition of oil well steel (N80) in simulated hydrogen sulphide environment by ferrous gluconate and synthetic magnetite. NAFTA 61:239–246

60. Okafor PC, Liu CB, Zhu YJ, Zheng YG (2011) Corrosion and corrosion inhibition behaviour of N80 and P110 carbon steels in CO_2 saturated simulated formation water by rosin amide imidazoline. Ind Engng and Chem Res. 50:7273–7281

61. Miksic BM, Kharshan MA, Furman AY (2005) Proceeding of 10th European Symposium of Corrosion and scale inhibitors

62. Kharshan M, Furman A (1998) Incorporating vapor corrosion inhibitors (VCI) in oil and gas pipeline additive formulations NACE, Corrosion 98(236)

63. Muller S, Syed QA, Yokose K, Yang W, Jackel M (2009) Corrosion inhibitors SCUP Home

64. Houghton CJ, Westermark RV (1983) J Pet Technol 239

65. Hovsepian PE, Lewis DB, Munz WD, Lyon SB, Tomlinson M (1999) Combined cathodic arc/unbalanced magnetron growth CrN/NbN super lattice coatings for corrosion resistant applications.

Surf Coat Tech 120(121):535–541

66. http://www.europipe.com

67. Westood J (2011) Macro factors driving the global oil and gas industry and the subsea pipelines sector Toronto

68. Shiwei WG, Gritis N, Jackson A, Singh P (2005) Advanced onshore and offshore pipeline coating technologies Shangai, China: 2005 China international oil and gas technology conference and expo

69. Morgan JH (1987) Cathodic protection New York: McGraw Hill

70. Guyer JP (2009) An introduction to cathodic protection New York: Continuing Education and Development Inc.

71. Lazzari L, Pedeferri P (2006) Cathodic protection New York: McGraw Hill

72. Baeckmann WV (1997) Handbook of cathodic corrosion protection New York: McGraw Hill

73. Laoun B, Niboucha K, Serir L (2008) Cathodic protection of a buried pipeline by solar energy. Revue des Energies Renouvelables 12(1):99–104

74. Benathmane R (2003) Study and simulation of cathodic protection by impressed current protection of a buried work Blida: Department de Chimie Industrielle, Université Sâad Dahleb

75. Canadian Association of Petroleum Producers (2009) Best management practices for the mitigation of internal corrosion in oil effluent pipeline systems Calgary: Annual report. CAPP

76. NACE (2003) Petroleum and natural gas industries–materials for use in H2S containing environments in oil and gas production–part 3: cracking-resistant CRAs (corrosion resistant alloys) and other alloys Houston: NACE

77. Bhavsar RB, Hibner EL (1996) Evaluation of corrosion testing techniques for selection of corrosion resistant alloys for sour gas service Houston: NACE International

78. NACE (2002) Corrosion costs and preventive strategies in the United States Houston: NACE

Chapter 6

Governing Shipping Externalities: Baltic Ports in the Process of SOx Emission Reduction

Daria Gritsenko[1] and Johanna Yliskylä-Peuralahti[2]

[1]Aleksanteri Institute, University of Helsinki, (Unioninkatu 33), 00014, Helsinki, Finland

[2]Center for Maritime Studies, University of Turku, Joukahaisenkatu 3-5B, 20014, Turku, Finland

ABSTRACT

This paper analyses the debate which has unfolded in the Baltic Sea Region regarding the reduction of sulphur content in vessel fuels, in order to illustrate how tightening environmental regulation challenges traditional forms of maritime governance. Using an interactive governance approach, this study reconstructs the process of sulphur emission reduction as a complex multi-stakeholder interaction in multiple contexts. The empirical investigation has drawn on documentary material from around the Baltic region, including Russia, and has applied the method of qualitative content analysis. The empirical study focuses on two interlinked questions: (1) How sulphur emission reduction policies are being anticipated by maritime industry, in particular by Baltic ports and (2) How port adaptation strategies are tied into Baltic local and energy contexts. Addressing these questions highlights the role of polycentricity in shipping governance and explains how the same universal international regulations can produce varying patterns of governance. The paper concludes that policy-making shall take an account of the fact that the globalized shipping industry is nevertheless locally and sectorally embedded.

INTRODUCTION

Environmental concerns have a growing strategic importance for maritime transport and ports (Darbra et al. 2009; Hall and Jacobs 2010). Even though shipping is a late-comer in environmental matters compared to many other industries, the tightening regulation and pressures from the end customers to diminish the environmental burden of the entire transport chain have changed the mind-set of the maritime transport sector (Lai et al. 2011). This paper focuses on air emissions of shipping, in particular the debate regarding reducing the sulphur content in vessel fuels to illustrate how tightening environmental regulation challenges traditional maritime governance forms. Even though vessels' air emissions are not the only negative environmental externality of commercial shipping, the rising energy costs, the need to cut harmful emissions to the environment and to increase the energy efficiency of the vessels has put the air emissions and energy question in the spotlight. The reduction of air emissions from shipping has been a hot topic in the maritime industry for over a decade, but with the entry into force of the amendments to the International Convention for the Prevention of Pollution from Ships (MARPOL) Annex VI "Prevention of Air Pollution from Ships" in 2011, the European Union (EU) Directive 2012/33/EU of 21 November 2012 regulating the sulphur content of marine fuels and the designation of the Baltic and North Sea, as well as the English Channel to Sulphur Emission Control Areas (SECAs), its acuteness has become evident. The debate about controlling vessel emissions reflects a wider shift of power from flag to port states (Roe 2013), the changing role of the ports in global transport and value chains (Olivier and Slack 2006), and political tensions regarding energy security (Aalto 2008). Previous research has addressed the technical side of emission reduction, focusing mainly on vessels (Endresen et al.2003; Eyring et al. 2005; Corbett et al. 2009; Wahlström et al. 2006; Balland et al. 2013), whereas the socio-political side of emission reduction in shipping, in particular the role and stakes of actors and polycentricity have been investigated far less (Roe 20082009; Ostrom 2012; Bloor et al.2013). Moreover, emission reduction in shipping has not been treated as an energy policy issue, even though uncertainties regarding fuel supply, distribution infrastructure and prices of fuels form the core of the matter for the shipping industry. We use the case of sulphur oxides (SOx) emissions in the Baltic Sea Region (BSR) in order to put these issues under social scientific scrutiny.

The study aims at reconstructing the process of SOx emissions reduction in the BSR as a complex multi-stakeholder interaction by tracing the actors' engagement in SOx emissions reduction policies in multiple contexts by using interactive governance approach (Kooiman 2003; Kooiman et al. 2008). It focuses on two interlinked empirical research questions: (1) How SOx emission reduction policies are being anticipated by maritime industry, in particular by Baltic ports and (2) How port adaptation strategies are tied in with Baltic local and energy contexts. The empirical investigation draws upon documentary material and applies the method of qualitative content analysis (Gläser and Laudel 2009). The sources used consist of materials from around the Baltic region, including Russia.

Recent literature has emphasized the need for stronger empirical maritime governance, in particular focusing on uncovering the failure of hierarchical regulatory governance of shipping negative externalities and researching the practical impact of institutions (Ng et al. 2013). Acknowledging this need, this research aims to contribute to the discussion on maritime governance through a thorough empirical investigation which emphasizes why polycentricity needs to be taken into account when shipping is concerned. Taking into consideration the trans-boundary and globalized character of shipping industry, the paper aims to explain how the same universal international regulation produces varying governance patterns. The key argument is that globalized shipping industry is nevertheless locally and sectorally embedded.

The structure of the paper is as follows: "Interactive governance approach and maritime governance" section introduces the interactive governance approach, "Data and method" section presents data and methods, "Analyzing governance: empirical research strategy" section explains the analytical strategy and operationalisation of the governance approach, "Case-study: Baltic ports in vessel SOx emission reduction" section elaborates on the analysis and its results, "Baltic ports: adapting to a complex environment" section discusses the results in the light of interactive governance approach, and "Conclusions" section concludes.

Interactive Governance Approach and Maritime Governance

Governance is a complex notion which allows for multiple interpretations. It has been conceptualized and analyzed as interaction (Kooiman 2003), multi-dimensional policy process (Hill and Hupe 2002), networks (Rhodes 1996), choice and implementation of instruments (Lascoumes and Le Gales 2007), dynamics between the structure and process (Börzel and Risse 2010), and discourse (Hajer 1995). Here we follow the interactive governance approach as elaborated by Kooiman (19932003), which is based on the assumption that 'societies are governed by a combination of governing efforts' (Kooiman et al. 2008:2) which reflect the multiplicity of societal responses to modern challenges. The interactive governance approach focuses on the occurrence of governing interactions "at different societal scales, from the local to the global and with overlapping, cross-cutting authorities and responsibilities" (Kooiman et al. 2008:2). It views governance as a variety of horizontal networks, as well as vertical arrangements between public and private entities (Kooiman 2003; Kooiman et al. 2008; Torfing et al. 2012).

One of the key strengths of interactive governance is its attention to polycentricity, which raises questions of how power is being exercised by actors located at different places and jurisdictional levels, and how multiple centers of decision making independent of each other can coexist (Ostrom et al. 1961; Ostrom 2010). Polycentric institutional structures enable addressing environmental problems at multiple scales and in dynamic environments where adaption to change and uncertainty is crucial (Galaz et al. 2008; Moss 2012). In policy studies, an interactive governance approach has also been applied to clarify collaborative policymaking and participatory processes, "whereby government involves its citizens, social organisations, enterprises and other stakeholders in the early stages of public policymaking" (Edelenbos 19992004:111). Thematically, interactive governance has most widely been applied to the study of river and water management (e.g. Moss 2004; Jentoft 2007; Edelenbos et al. 2010).

The study of the maritime emission reduction debate within an interactive governance perspective draws the researcher's attention to (1) the multiplicity of actors, whose number, positions and other intrinsic characteristics are subject to empirical verification, rather than a theoretically predefined parameters; (2) the variety of governing interactions, where a set of allowable actions and their outcomes are constrained by conditions both internal and external to the process under scrutiny; (3) the role of both formal and informal institutions in structuring the process. An actor is defined as "a single individual or a group functioning as a corporate actor" (Ostrom 2011:12). In the process of social interaction, externalities may occur if some actors do not find it in their interest to take account of the consequences of their actions on others (Buchanan and Stubblebine1962). Essentially, externalities point out to divergence between private and social costs. Institutions are regarded as 'rules of the game', representing formal rules and compliance procedures as well as agreed operating practices (informal rules) that structure the relationships between actors and socio-economic structures (North 1990; Scharpf 1997; Ng et al. 2013). Institutions influence actors' decision making. The context is the set of regulatory, technological, social, political, environmental, territorial and economic conditions, in which the governance process takes place. Positionality refers to an actor's relational position to space and time (Sheppard 2002). The interactive governance approach thus offers a framework for grasping the complexity of socio-political interaction in the emission reduction debate by focusing on how new forms of governance combine with traditional institutions and policy processes and how private actors can position themselves to sustain their position and create added value in the process of governance restructuring. The analysis focuses on actors who are involved, their strategic interaction, the multiple interconnected contexts in which interactions are embedded, and how institutions structure the interaction.

Within maritime governance, the interactive governance approach is valuable in uncovering inconsistencies regarding interactive mechanisms, in particular by drawing attention to the interaction between the governing system and the subject-matter of governance (Kooiman 2003; Torfing et al. 2012). Tan (2006), Roe (20092013), Marsden and Rye (2010) have shown that traditional forms of maritime governance which rely on nested hierarchies are ill-suited to integrating environmental concerns. However, the creation of governance structures capable of reducing the negative impacts of maritime transport is not an easy process, as the emission regulation case of this paper demonstrates. In maritime transport, as in other multi-stakeholder environments, the regulation and management of environmental externalities often faces the ambiguities of governance architecture, creating situations in which small-scale problems are being approached from higher levels and attempts to tackle global problems are being undertaken at the lowest levels (Young 1994), thus creating problems of 'fit'. A mismatch between the geographical extent of the emissions to be reduced and the territorial scope of the several institutional layers regulating maritime transports is notable in the case of sulphur emissions. Furthermore, a misfit between governance process and structure, differing understandings of the emission problem, limitations of available solutions and the actors' capabilities to fulfill the requirements further complicate the situation.

Data and Method

This study consists of an analytical investigation of governance changes brought about when new practices emerge in the industry, facilitated by new forms of regulation and a growing demand for higher environmental quality. The study is based on an analysis of qualitative data, carried out in order to explain how the ongoing process of air emission reduction in shipping changes maritime governance, especially with regard to shifting the position and action strategies of the ports. The data used stems from multiple sources: conference and seminar presentations, publicly available policy documents, industry statements (open letters, resolutions, press releases), news and statutory documents. Complementary to the 'desktop study' (Wheeler and Peszynska 2002), primary data was generated by the other author via participant observation and personal communication with maritime stakeholders in three events held in 2011–2012: 1) the Baltic Ports Conference Rostock, Germany 2011 (Baltic Ports Organization 2011), 2) the PENTA project workshop 'Major shifts or business as usual? Implications of sulphur regulation on maritime transport' held at Muuga, Estonia on 18.4.2012 (PENTA 2012), and 3) the CLEANSHIP Mid-Term Conference in Riga, Latvia 2012 (CLEANSHIP 2012). Various presentations in these events dealt with the policy process of the SOx regulation, estimates of how emission regulation will affect maritime transport flows and technology and how ships can comply with the regulations. By observing the discussions it was possible to obtain information on the reactions of the ports towards the emission regulation. This data in the form of field notes was used to complement written sources. In addition to the published and unpublished primary material, this study draws upon previous studies. Different types of data were nevertheless collected simultaneously (2012–2013). The mixed character of the data is a part of the complementary research design implemented in this study. Since any given type of data can give only certain kind of information, a combination of different types of data allows them to complement each other (Brewer and Hunter 2006; Small 2011).

The qualitative analysis of the content was chosen as the central data analysis method and was applied coherently to all primary data generated in course of research (Gläser and Laudel 2009). The primary aim of the analysis is to look for mechanisms which can account for unfolding events observed in the data, and to clarify if the propositions created on the basis of governance theory are mirrored in stakeholder debates; i.e. if other actors and the ports themselves acknowledge their changing role in maritime governance. Technically, this method of analysis consists of the two distinct steps: data compression and pattern recognition. Firstly, the data collected for analysis is compressed in a tabular form in accordance with categories which can be viewed as 'containers' for meanings, deductively derived from the theory. Though the extraction process is theory-guided, it remains open to the new concepts emerging from the data. The dimensions of potential interaction to be explored are the subject of interaction with the nature of the situation and the scope of the included actors. In the extraction tables, the information is summarized for theoretical reasons, so that the background cases (units of observation) are left in the background, whereas the information is preserved. Information with the same meaning is aggregated, whereas contradictory information is kept for further in-depth investigation. Thus, single units of analysis data are aggregated into larger units (referred to as variables or categories) at a more abstract-theoretical level, which allows subsequent analysis. During the analysis of extracted information, attention is paid to patterns in the data, in particular (1) Sequences of events that occur more than once, (2) Combinations of conditions, processes or outcomes that occur more than once; (3) Conflicting accounts of events or processes. Pattern recognition is thus the second step of the analysis. Once patterns are identified, the results of the analysis can be visualized and presented in form of graphical displays, which are used to both describe and explain the patterns in the data (Miles and Huberman 1994). The extensive use of graphical displays firstly allows for the reduction of the data by selecting and transforming it through summarizing the argumentation lines, and secondly, reinforces the validity of research by assembling information contained in lengthy qualitative data in accordance with clear-cut rules (Miles and Huberman 1994:11). Finally, the construction of displays plays a substantial role in deductive qualitative analysis, not only by helping to organize the material, but also by adding an analytical level to the interpretation of the data and by transferring it a certain level of abstraction.

Analyzing Governance: Empirical Research Strategy

The interactive governance approach offers a broad and inclusive framework well-suited to grasping the complexity of contemporary socio-political interaction in the field of public policy. The interactive approach to governance provides a researcher with the following conceptual guidelines: encouragement to look for actors and focus on their interactions in multiple institutional contexts (in which the interactions are embedded), as well as paying attention to how these contexts are interconnected (Kooiman 2003). The primary focus of the empirical investigation is on action, in that the data sources were used to extract information on the actors and the institutional contexts in which they interact, as well as the relationships between the actors and the contexts. Key questions included: Which actions are pursued, announced, desired, rejected, and intended? How is the modus of action embedded in contexts?

For the purpose of empirical investigation, the theoretical concepts introduced in "Interactive governance approach and maritime governance" section were operationalised in the following way. An actor is described in two dimensions: type (individual, institution, public, business etc.) and position, in regard to governing interaction (rules demanders, suppliers and targets). In relation to action, the category of actor is seen as empirically opened up. Thus, any actor who acted in regard to the policy problem was regarded as relevant. Institutions are operationalised with the grammar of institutions developed by Crawford and Ostrom (1995). They distinguish between three types of institutions: rules, the most prescriptive and formal type of an institution; norms, which are also prescriptive but are not presupposing any formal punishment for non-compliance; and strategies, which do not contain normativity but rather describe common communication patterns. Institutional statements can be described in three dimensions: the subject of institutional statement (what is regulated), its content (how it is regulated), and scope (to whom it is applied). The elaboration of these three dimensions allows for depicting the interconnectedness of actors and institutions by showing how spatial and institutional contexts, related to regulatory, technological, social, political, economical, and environmental elements, affect actors' behaviour. Institutional contexts exist simultaneously and overlap, thereby enabling some actions and putting constraints on others. Eventually, in relation to action, institutional contexts are characterized by two dimensions - opportunity and challenge - which describe the perceived influence of the 'rules of the game' on the potential outcomes of the process.

In order to explain how Baltic ports are affected by the new regulations and how they adapt to these changes, the analysis cannot be reduced to the simplistic scheme of 'regulator → regulated', but must embrace the complexity of the process. The use of interactive governance as a research framework has several advantages in this respect. Firstly, it focuses on the empirical research into dynamic action (governing interaction) rather than static structures (government institutions). Secondly, it makes no normative assumptions about the actors, so that the scope of governing actors is defined empirically for each concrete case. Thirdly, it does not impose any methodological restrictions, being open to quantitative, formal and qualitative contributions alike. Finally, although the policy process in question can be formally regarded as a new regulation which targets vessels and requires them to technically adjust to the new sulphur exhaust limits, the interactive approach emphasizes the governance dimension of this process. At the same time, this analytical strategy is not without limitations. The concepts of the interactive governance approach offer a rather simplistic research model to deliver an analysis of this process, because in the actual process of SOx emissions reductions, all the dimensions of governance are interconnected and influence each other so that the causes and effects are often difficult to distinguish between. Thus no tools are provided for recognising the direction of causality and the actor-structure problem remains unresolved.

Case-study: Baltic Ports in Vessel SOx Emission Reduction

Sulphur Reduction: Governing Actors

The mechanisms associated with SOx emissions reduction, such as monitoring, certification, control and penalisation, as well as the provision of additional services, relies on multiple actors with different and often conflicting agendas. The logic of the interactive governance approach suggests that the question "Who is the governing actor?" is empirically open. The initial monitoring of the data showed that both public and private bodies consider their interests affected and therefore are concerned with the upcoming changes in shipping SOx emissions standards (Table 1). Additionally, the exploration focuses upon the actor's position in governance interaction, distinguishing between rule-demanders, rule- suppliers and targets of rules (Buethe 2010). This framework allows for a systematic approach to identification of positions of governing actors. Importantly, it highlights the changing relationship between actors in the development of governance processes.

Table 1: Actors and their roles in sulphur emission reduction process (own compilation)

Actor	Function/regulatory domain/ governance role
Inter-governmental organizations: International Maritime Organization (IMO), Helsinki Commission (HELCOM)	Resolutions and recommendations
The European Union (EU), European Maritime Safety Agency (EMSA)	Sulphur directive 1999/32, clean shipping projects, enforcement, monitoring and control
Nation states	Acting as 'flag', 'port' and 'coastal' states in terms of regulatory and implementation capacity
Private regulators (marine insurers, P&I clubs, classification societies)	Monitoring and ensuring that the regulation in place is obeyed
Ship owners and their industry associations (e.g. ECSA & associations at national level)	Targets of the rules, increasing demanders of change and rule-suppliers (e.g. Clean Ship Project)
Cargo-owners (consignors) and associations representing their interests	Actors in supply chain, increasingly rule-demanders
Ports and their associations (European Sea Port Organization, Baltic Port Organization)	Mixed role both as a market player and potential facilitator
Shipbuilding industry, manufacturers and suppliers of scrubber technology	Market interests at stake
Producers & suppliers of fuels, bunkering companies	Market interests at stake
NGOs and civil society	Environmental pressure – not very strong in the case of SOx emissions reduction

Gritsenko and Yliskylä-Peuralahti

Gritsenko and Yliskylä-Peuralahti *Maritime Studies* 2013 12:10, doi:10.1186/2212-9790-12-10

Nation states are usually considered as the main rule-demanders, since they have representatives in the intergovernmental organizations (such as the International Maritime Organisation - IMO), through which they seek to formally regulate the negative environmental effects of shipping. In the Baltic Sea region, the Helsinki Commission (HELCOM) has a key role in environmental protection and policymaking through intergovernmental cooperation (Stankiewicz 2012), although bringing the divergent interests of the Baltic littoral states together is not always possible (Knutsen and Hassler 2011; Hassler 2011). The EU institutions (the European Commission and the European Parliament) are now increasingly active in demanding new environmental regulations as well as attempting to introduce rules for exceeding the level of demands posed by the states in the IMO debates. In EU policymaking, issues related to public health and the environment are lobbied for by the interest groups through the Directorate General Environment, although there are often conflicts of interest with the maritime and shipping industry, as represented through the Directorate General for Mobility and Transport (Pallis 2007). Thus, both intergovernmental organizations and EU institutions are arenas in which different interests within maritime transport meet each other. At the same time, the role of environmental NGOs, which played a prominent role in demanding new rules and regulations in other issue-areas, has been rather marginal in the maritime issues (Skovgaard 2012), especially concerning the SOx questions.

The private actors can be expected to occupy the target-of-rules position in governance. In the Baltic Sea region, ship and cargo-owners and associations representing their interests continually complain about a sharp rise in vessel operation costs caused by the sulphur regulations (Sweco2012; field notes). Also in the bunkering sector, especially in Russia, SOx regulation was received most negatively, due to fears of significant shrinkage of the market and loss of market shares (PortNews 2013). However, the Baltic maritime industry is not homogeneous in its reactions to SOx regulation. A small group of environmentally-minded ship owners (e.g. Wallenius Wilhelmsen, Maersk) have already adopted voluntary measures to implement new regulations in order to gain a positive image (Maersk 2011; Wallenius 2013). Others have started to invest in adaptation strategies well in advance; e.g. by transferring their vessels to liquefied natural gas (LNG) (Izvestiya 2012). Demand for additional measures comes also from a group of cargo-owners willing to transport their goods with a vessel which adheres to environmental quality standards. Since documentation from a classification society relates only to upholding the minimum standards, ship owners turn to ISO certification or private certification schemes to prove their environmental quality. In the BSR, one of the prominent voluntary certification schemes used for notifying environmental quality which goes beyond regulation is the Clean Shipping Index (CSI) (CSP 2012; Wuisan et al. 2012). The CSI was created to improve the environmental performance of maritime transport and to bring together like-minded vessel and cargo owners. After fulfilling the requirements of the Index, the vessel-owners receive authorization for their vessels from a classification society. Cargo-owners in turn can use the CSI members in their procurement of transport services, to assure their own customers that they only use companies with good environmental performance. In addition, a Clean Shipping Forum, which other supply chain actors such as freight forwarders and ports can join, has been created. Engaging in such schemes as CSI, ship and cargo owners change their position with regard to governance interaction.

Multiple Contexts of the Emission Reduction Process

The interactive governance framework suggests that the institutional context shapes the governing interactions. Thus, it is beneficial to study how actors relate to the context. In order to analyzing the assumed actions and intentions of the actors, we relied on a broad desk-top study of publicly available materials to reconstruct the broad lines of the multiple contexts in which the debate on SOx unfolds. At this stage, qualitative content analysis allowed for identifying how the actors relate to the context and how the particularities of the context shape actors cost-and-benefit calculations with regard to the expected outcomes. When the research was consolidated, we recognized a need to enlarge the contextual analysis beyond the shipping arena and consider the regional governance patterns. In particular, we pinpointed the differences in debates on the SOx emissions which unfolded among the EU and Russian maritime stakeholders. The context of energy value chains and conflicting interests in the field of energy policy appeared very strong through the source materials.

Regulatory Context

The IMO and the EU, the rule-suppliers regarding SOx emissions reduction, have moved forward with the issue with a different timetable and on a different geographical scale. The IMO Marine Environment Protection Committee adopted amendments to the MARPOL Annex VI regulations on sulphur oxide requiring a maximum 3.50% content by 1 January 2012 and 0.50%, by 1 January 2020 globally. As coming to an agreement on global SOx levels turned out problematic, sulphur emission control areas (SECAs) were formed. These consisted of: Baltic and North Sea, North American SECA and Japan-East Asia SECA. The transposition of provisions set under the IMO into the system of EU law has had important consequences. Firstly, the existence of an EU directive gives the Commission a degree of competence in this area, which means that the member States are supposed to give an agreed position at the IMO for all matters concerning the sulphur content in liquid fuels. Secondly, a directive is the most prescriptive instrument in the EU legislative tool-kit. It has the features of direct applicability and direct effect, making the implementation and enforcement more detailed and rigorous than the provisions of international law. Moreover, this directive is applicable on a territorial basis, so that any foreign vessel (i.e. not sailing one of the EU Member States flags) also falls under the scope of its provisions when entering the coastal waters and ports of the EU countries. Additionally, the geography of the Baltic Sea and its shipping routes make it practically impossible for vessels to avoid sailing in EU coastal waters, even if their target ports are located in Russia. Thus, the EU sulphur directive *de facto* has a wider scope (of actors included). Moreover, in the IMO, practices of lobbying for delays, individual opt-outs and special schedules are widespread, although the EU Commission has refused to revise the directive's schedule. Finally, EU law adds a private liability dimension to IMO rules, whereas private entities cannot be penalized for non-compliance with IMO public international law. Furthermore, the EU decided to take additional actions and in its transposition set even tighter limits. The EU directive 1999/32 limited the ship fuel sulphur content to 0.1% in 2015 in the Baltic Sea and North Sea SECA area. Non-compliance with the directive is to be penalized by the national authorities. However, checks on the amount of fuel in ships were made very loosely, which raised concerns among NGOs that the failure to comply will not always be discovered (Transport and Environment. Newsroom 2012)

These legal inconsistencies have direct implications for the functioning of ports. In compliance with MARPOL Annex VI a regulatory system based upon Air Pollution Prevention Certificates emerged. The Certificates are applied by shipping firms owning and/or operating vessels from respective national agencies (e.g. maritime administration), which verifies the compliance of vessels of its nationality with measures falling under the scope of air pollution reduction (engine modifications, supply of low-sulphur bunker etc.). Neither the IMO nor the EU regulations lay down detailed provisions on sulphur legislation enforcement. The states are responsible both as flag and port states for the successful implementation of the regulation. Thus, the role of inspector is reserved for ports, which in the framework of port State control (PSC) procedures verifies the vessel's compliance based on the vessel's certificate as prima face evidence. If a vessel is found to be non-compliant, compliance and enforcement mechanisms (e.g. reporting deficiencies, monetary penalties, the removal of certificates or documents of compliance) can be triggered. A similar situation emerges if a vessel chooses to adhere to the IMO sulphur regulations by installing a scrubber, which requires a retrofit plan, with deadlines and approval by class or flag state. Classification societies need to inspect a vessel and state whether the scrubber is fully functional. Thereafter, ports come into action, since the scrubber installation is monitored by PSC authorities during a PSC inspection. Therefore, controlling function of ports in respect to monitoring the implementation of sulphur fuel reduction (inspecting, sampling, prosecuting) is central, even though the regulatory environment is ambiguous in respect to implementation of this function.

"Environment vs. Economy" Context

Even though air emissions represent only a part of the negative shipping externalities, the issue is significant for several reasons. The shipping industry accounts for approximately 10-15% of NOx (nitrogen oxides) and 4-9% of SOx emissions globally (Endresen et al. 2008). Sulphur in ship fuels lead to declining air quality, and creates negative health effects for humans and unwanted acidic loads in the environment. Sulphur in fuels is transformed into sulphur oxides after a burning process producing the key component of the particle matters (PM). Therefore, the regulation of the SOx content in a bunker is an effective way to contribute to limiting the adverse health effects of ship exhausts (Jalkanen et al. 2009). The concerns related to the negative effects of sulphur emissions on public health and the deterioration of the marine and coastal environment were, for the regional maritime stakeholders, the main reason for bringing the vessel fuel issue into the realm of public policy. At the same time, many indicated that the decision to introduce regulation was premature and warned of the potential negative environmental effects connected to modal shifts from sea to rail and road (field notes).

The choices regarding fuel vessel use is a significant driving force for changing future transport flows because fuel costs represents between 45 to 55% of the daily operational costs of vessels. The proposed SOx regulations are expected to raise the costs of shipping between 30 to 50% (Kalli et al. 2009; EMSA 2010). Ship and cargo-owners and their associations point out that, as a result of SOx regulations, freight prices will increase; since additional costs will be allocated to charter parties. Export industries particularly fear a significant increase in sea freight prices (with up to 1.2 billion Euro additional costs per year), which in turn can negatively affect their competitiveness. Many industry associations have lobbied to postpone the introduction of stricter sulphur emission standards by 5 to 10 years (Sweco 2012; Forest Industries, Swedish Forest Industries Federation2012; EK 2011). Ship and cargo owners expressed objections towards the EU Sulphur Directive 2012/33/EU because it sets tighter limits for vessels operating inside the IMO SECA areas as compared to other waters under the jurisdiction of EU member states, creating an uneven playing field within the EU. They proposed that either another SECA should be formed to cover the Mediterranean Sea, or that, alternatively, compensation measures should be offered for those vessels operating inside the presently established SECAs in order to equalize the differences in operation costs (CEPI 2012; Maritime 2012; Yle Uutiset 2012).

Also, in the bunkering sector, serious concerns were shown, in particular by the Russian bunkering industry. In 2011 the Russian Association of Marine and River Bunkerers held an annual forum, in which a petition was prepared and later sent to the Russian Prime Minister Medvedev, in order to draw the attention of the federal government to 'Problem 2015' (SRO RAMRB 2012). The main concerns of the bunkering industry are (1) the inability to supply a sufficient amount of compliant fuel due to lack of necessary refinement capabilities in the Russian oil industry; (2) the shrinkage of the bunker market, particularly of the transit bunkering segment; (3) the absence of standards, equipment and infrastructure for a wide introduction of LNG bunkering in Russian ports. The main competitive advantage of the Russian bunkering market has been lower prices for heavy fuel oil (HFO), which created a constant growth of ca. 10% due to a gradual increase in transit bunkering. As regards new sulphur regulations, the bunkering industry appealed to both rising costs connected to adjustment (modernization of old and acquisition of new equipment for refining and bunkering), decreasing incomes connected to loss of competitiveness with foreign companies, and a drop in number of vessels bunkering in Russian ports. In the worst case scenario, the prognosis for Russian bunkering market shrinkage is up to 40% (Morskoy 2012b; Morskoy 2012d). The Russian ship owners fear that they will have to use more expensive low-sulphur fuel due to the absence of sufficient LNG infrastructure (Morskoy 2012b). At the same time, the Bunkering association recognized the perspective of LNG bunkering development as a long-term opportunity.

Finally, due to changes in the operational environment (rising freight prices, modal shift, fuel shortages, lack of infrastructure), the traffic patterns in many ports are expected to change. In particular, the small and middle-sized ports positioned within the Baltic SECA are afraid of losing traffic, their main source of earnings, because growing maritime transportation costs could increase cargo transports on land or cause the re-routing of cargo to ports outside the SECA area (PortStrategy 2011). Russian ports which now enjoy a higher number of transit ship calls visiting for bunkering purposes are also worried about losing their traffic (PortNews 2013). As ports with bigger volumes are in a better position in terms of market shares and are more attractive for investments, from the smaller ports' perspective the new tighter environmental regulation will only enforce the on-going trend towards larger ports by eradicating the small ports.

Technological Context

Currently there are three technical ways a vessel can fulfill the proposed emission limits: (1) by using fuels with a reduced sulphur content (marine gas oil (MGO) or marine diesel oil (MDO)), (2) installation of a scrubber (exhaust gas cleaning system) to a vessel, or (3) conversion of a vessel so that the use of liquefied natural gas (LNG) or other alternative fuels is possible (Kalli et al.2009; EMSA 2010; Bengtson et al. 2011; Stenhede 2012). However, all these options have technical and/or financial limitations. MGO/MDO is already available, but many ship owners restrain from using it due to its higher price. Additionally when NOx standards come into force, the ship owners who chose this alternative will need a new round of adjustment measures. With a scrubber, a vessel can continue to use high sulphur bunker, but a scrubber costs several million euros and installation is not technically possible for all vessel types. Scrubbers also produce sludge and washing waters that cannot be put into the Balticsea. Ports thus need to have reception facilities for this material from vessels (Kalli et al. 2009; Wärtsilä 2010). Even though the suppliers of the scrubbers are very optimistic regarding the effectiveness of their technology, in the opinion of the ship owners/ operators the scrubbers currently offered on the market are not reliable enough to meet the emission standards (field notes CLEANSHIP 2012: Hoenders' verbal communication 20.9.2012, Morskoy2012a). The use of LNG can solve many environmental problems, as it naturally contains no sulphur and the emission caused by LNG used as a fuel are lower compared to other fuels (EMSA 2010). However, LNG is not a panacea either: vessels need to be converted and terminals suitable for LNG bunkering are required. Also the lack of operational standards produces multiple safety, security and environmental concerns, hindering LNG usage (Aagesen 2012). Thus, what investments into infrastructure and facilities will be needed in terms of port adaptation is yet unclear. Shipbuilding industry, manufacturers and suppliers of scrubber technology, as well as producers and suppliers of new marine fuels all have the knowledge to support emission reductions from the technical side, i.e. they possess the expertise necessary to create realistic and efficient rules. At the same time, these actors all have market interests at stake: new SOx standards mean increasing business opportunities for them – and increasing competition.

Whereas the introduction of SOx regulation brings the ports new responsibilities as rule-suppliers since the responsibility to monitor and control compliance is held by the port state authorities, it also makes ports to a target for rules about infra- and superstructure development and maintenance. Ports become responsible for enabling cleaner operations via the establishment and maintenance of reception facilities (for scrubber wastewaters and sludge), LNG infrastructure (storages, bunkering terminals) and shore-side electricity facilities. For many Baltic ports and their associations, these changes constitute a major concern (Holma and Kajander 2012), since long-term investment decisions in which ports are dependent on the choices of other actors need to be made.

Regional Governance Context

The regional governance context adds an important dimension to our understanding of the Baltic ports' operational environment. The SOx emission reduction has an immediate connection with the energy policy agenda, since it directly concerns energy efficiency, the choice of fuels and their supply. In particular, the question of fuel availability and price poses challenges to ship owners, who have to adjust their vessels and ensure compliance with new exhaust standards. Which adjustment alternative will turn out to be the best economic solution depends on the disposition of forces in Baltic energy markets. So, besides the economic costs, the political costs of adaptation need to be accounted for.

In the Baltic Sea Region, energy questions are high on the political agenda. Thus, the patterns of Baltic regional governance necessarily influence the port adaptation strategies in the SOx emission reduction process. The recent development of the system of Baltic maritime governance was marked by EU attempt to widen its influence in international maritime affairs (Gritsenko 2013). The proliferating EU competences (as in case with the Sulphur Directive) put limitations on the strategies available to the Russian maritime industry. In particular, the Russian bunkering market is concerned about consequences of SOx regulation.

Each of the technological options required to meet the SOx requirements has a different meaning for the bunkering market. If the ship owners choose the MDO/MGO option, the bunkering companies need to increase their supply of these fuels. Bunkering constitutes 0.3-0.5% of Russian GDP. Nearly 80% of the fuel supply is HFO, which for some companies consists of 90-95% of their portfolio (Morskoy 2012b; Morskoy 2012d). Although up to 70% of the fuels the Russian bunkering companies supply to the BSR are bunkers with low-sulphur content (1%), this fuel will not be useable when the 0.1% limit comes into force in 2015. Currently these companies do not offer any fuel which is near the 0.1% limit, so a switch to MGO/MDO is unavoidable. The majority, ca. 70% (Morskoy 2012b) of Russian bunker suppliers are daughter companies of vertically integrated oil and gas companies, including Rosneft and RN-Bunker, Gazprom and GazpromNeft Marine Bunker, Lukoil and LUKOIL-Bunker, as well as the pipeline companies Transneft and Transneft-Servis. The rest consist of smaller independent bunkering companies (e.g. Baltic Fuel Company, Baltiyskaya Toplivnaya Kompaniya) which buy fuel from oil majors instead of producing it themselves. Thus, the dominant Russian bunkering companies, which are bound to their own production schedules, have started to produce marine diesel oil, which requires costly investment in the refinery and production processes. Eventually, maritime stakeholders assume upcoming modal shifts due to shortage of compliant fuels and/or rising fuel prices in Russian ports.

In case scrubber technology is chosen, Russian bunker companies and the oil industry can proceed with a 'business as usual' scenario, in which they are capable of supplying HFO at competitive prices. However, they cannot assess the scrubber option as the 'first-choice' alternative due to multiple technical issues which still remain unsolved (Izvestiya 2012). As for the LNG option, the opinions in Russian maritime sectors are divided: some see it as the future alternative (Izvestiya2012; Morskoy 2012a), some are sceptical due to the required infrastructure investment (Morskoy2012ab; Morskaya 2011). The EU is widely supporting and developing the idea of LNG-fueled European fleet (European Commission 2013). Russian gas supply capabilities are excellent, so the interests of the EU and Russian LNG suppliers meet. However, extensive use of LNG requires development of (1) a regulatory basis for LNG use; (2) new building and retrofitting; (3) LNG infrastructure (production, storage, bunkering terminals etc.) which are all costly procedures and will take a long time to implement (Morskoy 2012c SRO RAMRB 2012; Morskoy 2012b). In this case, the EU will have a competitive advantage. Statoil is leading the Nordic market for LNG bunkering, while many liquefying facilities are situated in Europe, so that Russian gas transported via pipelines can be liquefied and supplied as fuel by European producers. Since the Russian gas industry is monopolized by the state-owned Gazprom, Russian ports are afraid of losing their traffic in case the federal state will not help them to adjust, modernize and develop costly LNG infrastructure (Morskoy 2012c). Another solution for Russian ports would be to engage in closer cooperation with the EU partners to ensure sound development of LNG facilities. The economic viability of this solution can be undermined by the political situation, since it touches upon the sensitive field of EU-Russian energy relations. Thus, the advances in Baltic maritime governance are closely bound to the broader agenda of EU-RF relations (Gritsenko 2013).

Summary

Though SOx regulation targets vessels, our analysis has shown that, in the emission reduction process, the role of the Baltic ports will change. The ports will gain new controlling functions, but new operational requirements will also be imposed. Stevedore companies, ship and cargo owners, port service providers, bunkering companies and other actors interact and carry out their operations within a port. On behalf of the ports, the tightening of environmental regulations is a threat to competitiveness. Before a port can become an enabling environment for environmental shipping, the required port infrastructure has to be built. In order to finance such changes, ship-owners (via duties and fees) can be involved, or ports can invest their own resources or find investors among the third parties (nation states, EU funds, private investors). However, whereas larger ports might have more resources to pay for new facilities and better opportunities to manage co-financed schemes, smaller ports often lack resources and feel that their competitiveness is threatened. Many operators of small and medium sized ports are thus worried, because the operational costs of shipping are constantly increasing, while earnings are not (Stubbe2012).

Whereas some ports have concerns about their survival after 2015 when the IMO sulphur regulation enters into force, and are trying to calculate what it would mean for their traffic volumes, other ports in the BSR have actively started looking for available options. Examples include collaborative agreements between ports to ensure cleaner shipping. The Ports of Stockholm and the Port of Helsinki have cooperated in environmental issues since 2009. The Ports of Stockholm and the Port of Turku signed a collaborative agreement in 2011 to improve the environment in the Baltic Sea. The ports collaborate in facilitating the use of LNG for vessel fuel and for enabling LNG bunkering in both ports, by investigating the possibilities of supply electricity from the shore to more vessels operating with frequent liner schedules, and the management of grey and black water (The Ports of Stockholm 2011). All three ports (Stockholm, Helsinki and Turku) share the same shipping lines. By collaborating with the shipping line, the ports made preparations for the new vessel 'Grace', which uses LNG. The vessel started sailing on the Turku-Stockholm route in January 2013 (Viking Line 2013ab). Other examples include the establishment of collaborative networks. Currently, nine Baltic ports (Aarhus, Helsingborg, Helsinki, Malmö-Copenhagen, Tallinn, Turku, Stockholm, Sczcecin-Zwinoujscie and Riga) together with ship owners, LNG companies, national port organisations and European Seaports Organisation (ESPO) work together to enable LNG bunkering for vessels in the Baltic ports (TEN-T EA, Trans-European Transport Network Executive Agency 2012).This project highlights the facilitating role of ports in shipping governance process: the ports facilitate the investment, give land and make the necessary investments for the vessels (e.g. new quays). The port authorities also make a concession with a gas infrastructure company/gas operator to build and operate the LNG terminal. The project is TEN-T funded (Oldakowski 2012).

New shipping environmental standards can also bring business opportunities to ports. By virtue of their connectivity function, ports occupy a unique position precisely due to the fact that they constitute governance arenas which are inhabited by a large number of actors involved in transport and supply chains. This opens up possibilities of stewardship, new leadership and reputational gains. While tightening environmental regulation is currently seen as a threat to connectivity, it can also be a source of competitive advantage. Some shipping companies and cargo owners have already realized this, with regard to the increasing adoption of voluntary regulations such as the Clean Shipping Index (CSP 2012; Wuisan et al. 2012). Broadening the range of stakeholders and explicating the responsible parties (not only in terms of liability, but also in terms of responsibility) would change maritime transport and deliver clean and efficient solutions within the global supply chain (Ponte and Gibbon 2006; Lloyd's Register 2011; Pike et al.2011).

Baltic Ports: Adapting to a Complex Environment

Ports cannot be considered as 'newcomers' in the domain of shipping externalities governance, given their role in monitoring and control granted by the PSC mechanism. However, the governance of shipping externalities is becoming a polycentric and multi-leveled process, which requires venues where actors can meet communicate and create structures in which different governance mechanisms can complement each other. This strengthens the role of ports as a place (Olivier and Slack 2006; Verhoeven 2010) which can naturally offer themselves as venues due to their physical properties (prominently as waterfronts and logistic nodes) as well as their environmental stewardship in the course of the greening of maritime sector. On the other hand, this requires ports to adapt new strategies, which will allow them to function in a situation of multiple conflicting contexts, which can create uncertainties in their operational environment.

The first empirical research questions asked how SOx emission reduction policies are being anticipated by maritime industry, in particular by the Baltic ports. On the basis of our analysis, we defined two strategies which Baltic ports are likely to adopt in order to adjust to upcoming changes: (1) preventing loss of traffic by creating compliance-friendly infrastructures (ports themselves bear the cost of adaptation or find external investment); (2) raising the attractiveness of shipping as environmentally-friendly transport (e.g., in the light of CSR, cost of adaptation is transferred on the other actors) (Figure 1, right hand side). These strategies are not universal, but constitute responses to the possible future scenarios (chains of events expected to happen) emerging in the process of multi-stakeholder interaction and shaping ports' operational environment. As demonstrated above, when pursuing these strategies, ports can act as environmental leaders to promote change, as coordinators of multi-stakeholder processes, as well as cooperating with other ports.

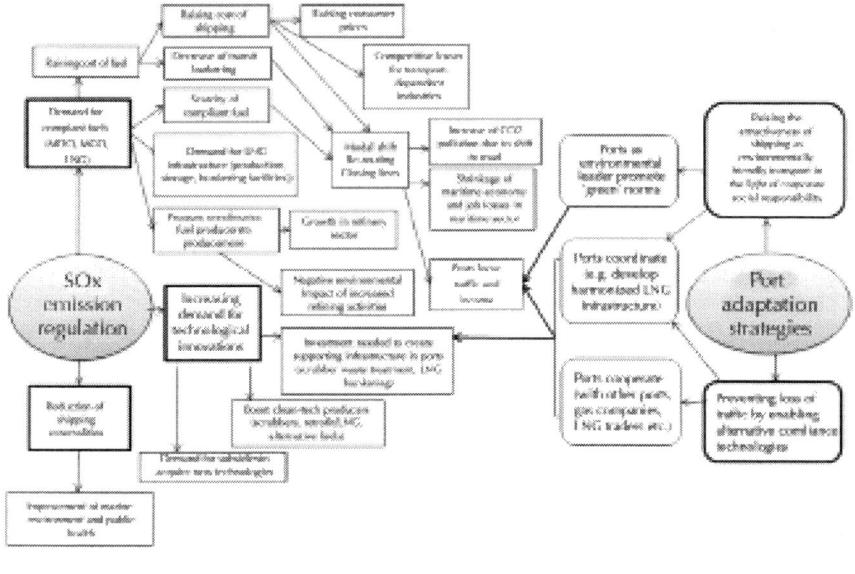

Figure 1: Baltic ports: adapting to a complex environment.

The 'mind-map' of the stakeholders' debate on the SOx reduction process presents their views of the process, mapping conflicting contexts in terms of opportunities and threats (Figure 1, left hand side). This brings together the analysis of governing actors, institutions, which structure their interaction, and conflicts between the different institutional contexts, within which the SOx debate unfolds. At the initial stage, most actors feared the upcoming changes. However, these changes do not necessarily have a negative impact on the maritime sector, or on the competitiveness and connectivity of ports. Changes occurring in ports, along with the proliferation of SOx emission reduction procedures, illustrate how these concerns are anticipated in the BSR.

The second empirical research question seeks to clarify how port adaptation strategies are tied in with Baltic local and energy contexts. At the same time, the strategies described above include benefits and challenges which are inherently connected to the broader socio-political context of the BSR. Within the Baltic Sea region, there is still an East–west divide regarding the actors' willingness and capacity to take on environmentally related investments. For example, public awareness about forming political attitudes and consumer pressure demands that such investments are higher in the Nordic countries and Germany compared to the Baltic States or Poland. Germany and Nordic countries also have longer history of implemented policies integrating environment and economic investments, and an indication of this is eco-innovation performance (Weber et al. 2012). Practically-oriented projects between ports could diminish this divide by facilitating best practices. Secondly, we need to account for energy issues: also emission reduction from shipping is a question of energy policy, which requires choice and changes in the energy markets. This means that the vulnerable energy policy balance in the BSR, which is dominated by the EU-RF relation in this area, also has to be taken into account.

As the Figure 1 indicates, the interaction of multiple stakeholders defines the structure of the ports' operational environment, which in turn influences the positions the ports assume in the governance process, and also defines their alternatives against the overall cost-opportunity structure. Moreover, the process is nested within the wider patterns of regional governance. The ports' adaptation strategies in the Baltic do not necessarily reflect the alternatives, which will be met by ports in other regions. Taking into consideration the transboundary and globalized character of the modern maritime industry, our results show how different governance patterns may emerge in different regions with regard to the same 'universal' international regulations. This again explains why the hierarchical regulatory governance of shipping through negative externalities fails and why polycentricity needs to be taken into account when shipping is concerned.

CONCLUSIONS

Using an interactive governance approach, this study has reconstructed the process of SOx emissions reduction in the Baltic Sea region as a complex multi-stakeholder interaction. In particular, it has paid attention to how governing interactions are embedded within local socio-political and energy contexts. The empirical investigation has demonstrated that the adaptation process is not limited to developing governance mechanisms capable of incorporating technical solutions in order to reduce negative impacts. Adapting to the changes in their operational environment, actors pursued new strategies, defined new rules and norms, but also reinterpreted the old institutions to reinforce their governance positions. This explains how seemingly universal international maritime regulations can produce varying patterns of governance.

Observing the process of adaptation of the Baltic ports, this article highlights the importance of how to create governance structures capable of reducing the institutional ambiguities and resolving the problems of collective action. The case study showed the major limitation of a 'technological fix' – its ignorance regarding the divergent interests of multiple actors at different levels and scales – which dismisses SOx emissions from any other policy arena rather than pollution reduction. However, the issue of low-sulphur shipping appears far more complex, since no single actor has the full authority and capability to implement it. The mechanisms which are activated to limit the SOx emissions from seagoing vessels, such as monitoring, certification, control and penalization, as well as service-provision, rely on multiple actors with different and conflicting agendas, which are disseminated within the polycentric shipping governance arena. Thus, the article claims that globalized shipping industry is inherently polycentric and that the local and sectoral embeddedness of these multiple centers of authority needs to be taken into account.

As the analysis through the interactive governance approach reveals, low-sulphur shipping challenges the traditional forms of port involvement in shipping governance, requiring changes in port administration and management. The introduction of SOx regulation broadens the ways in which of stakeholders are affected by the changes in shipping standards. However, this *de facto* change was not structurally anticipated by the introduction of comprehensive strategies for the management of shipping practices. Thus, many actors were forced into creating *ad hoc* and temporary solutions. Given the high degree of interdependence between the optimal strategies, the operational context has remained uncertain. The lack of clarity in terms of SOx adaptation strategies not only has economic but also political consequences, which have to be taken into account when planning how to transfer to zero-waste zero-emission shipping. In this study, we particularly suggest paying more attention to the position of ports in the process and their potential role in reducing the ambiguities and resolving the problems of collective action.

AUTHORS' CONTRIBUTIONS

The article is based on joint work of the authors DG and JYP. Both authors have read and approved the final manuscript.

ACKNOWLEDGEMENTS

This research was supported by the Finnish Doctoral Programme for Russian and East European Studies (Aleksanteri Institute/University of Helsinki) and University of Turku, Centre for Maritime Studies. The publication was granted by the Academy of Finland project "CHIP – clean shipping economics – shipping under the new paradigm" (decision number 257968) and project PENTAHLON – Ports of Stockholm, Helsinki, Tallinn, Turku and Naantali -together. The PENTATHLON project is financed by the Central Baltic INTERREG IV A Programme 2007-2013 of the European Union Regional Development Fund, the Ports of Stockholm, the Port of Helsinki, the Port of Turku, the Port of Naantali and the Estonian government. The article reflects the view of the authors. The Managing Authority of the INTERREG Central Baltic IV A Programme or the other oraginisations involved cannot be held liable for the information published in this article. The authors would like to thank two anonymous reviewers for their helpful comments.

REFERENCES

1. Aagesen, J. 2012. Lloyd's register LNG bunkering infrastructure study 2012. http://www.lr.org/sectors/marine/documents/237162-lng-as-fuel.aspx Accessed 17 Sep 2012 2008. The EU-Russian Energy Dialogue: Europe's Future Energy Security, Aalto P Aldershot: Ashgate.

2. Balland, O, SO Erikstad, K Fagerhult, and SW Wallace. 2013. Planning vessel air emission regulations compliance under uncertainty. Journal of Marine Science and Technology, doi:10.1007/s00773-013-0212-7

3. Baltic Ports Organization. 2011. Conference Rostock, Germany Sep 8–9, 2011, http://www.bpoports.com/477.html. Accessed 30 May, 2013

4.   Bengtson, S, K Andersson, and E Fridell. 2011. Life cycle assessment of marine fuels. A comparative study of four fossil fuels for marine propulsion. Chalmers University of technology. Department of shipping and marine technology. Report No 11:125. Gothenburg, Sweden. http://www.dma.dk/themes/LNGinfrastructureproject/Documents/Fuels%20and%20environment/LCA%20of%20four%20possible%20marine%20fuels.pdf Accessed 6 Jun 2012

5.   Bloor, M, H Samson, S Baker, D Walters, K Dahlgren, E Wadsworth, and P James. 2013. Room for manoeuvre? Regulatory compliance in the global shipping industry. Social & Legal Studies 22 (2): 171–189.

6.   Börzel, TA, and T Risse. 2010. Governance without a state: Can it work? Regulation and Governance 4 (2): 113–134.

7.   Brewer, J, and A Hunter. 2006. Foundations of multimethod research: Synthesizing styles. Thousand Oaks, CA: SAGE.

8.   Buchanan, JM, and W Stubblebine. 1962. Externalities. Economica, New Series 29 (116): 371–384.

9.   Buethe, T. 2010. Private regulation in the global economy: a (P) review. Business and Politics 12 (3): 1–38.

10. CEPI. 2012. Confederation of the European paper industries. Press release May 25, 2012. Low sulphur fuel directive: EU industry competitiveness again disregarded. http://www.paper.org.uk/news/2012/Press%20Release_Sulphurcontentagreement.pdf Accessed 17 Sep 2012

11. CLEANSHIP. 2012. Mid-Term Conference in Riga, Latvia 19.-20.9.2012. http://www.clean-baltic-sea-shipping.com/information/file/121 Accessed May 30th, 2013

12. Corbett, JJ, H Wang, and JJ Winebrake. 2009. The effectiveness and costs of speed reduction on emissions from international shipping. Transportation Research Part D 14 (8): 593–598.

13. Crawford, S, and E Ostrom. 1995. A grammar of institutions. American Political Science Review 89 (3): 582–600.

14. CSP. 2012. Clean Shipping Project. 2012. Clean Shipping Index. http://www.cleanshippingindex.com/about/   Accessed June 27th, 2013

15. Darbra, RM, N Pittam, KA Royston, JP Darbra, and H Journee. 2009. Survey on environmental monitoring requirements of European ports. Journal of Environmental management 90 (3): 1396–1403.

16. Edelenbos, J. 1999. Design and management of participatory public policy making. Public Management 1 (4): 569–578.

17. Edelenbos, J. 2004. Institutional implications of interactive governance: insights from Dutch practice. Governance 18 (1): 111–134.

18. Edelenbos, J, N Van Schie, and L Gerrits. 2010. Organizing interfaces between government institutions and interactive governance. Policy Science 43 (1): 73–94.

19. EK. 2011. Confederation of Finnish Industries. 15.7.2011. The revision of the sulphur directive is a disappointment. http://www. ek.fi/ek/en/news/the_revision_of_the_sulphur_directive_is_a_ disappointment-7707 . Accessed 6 Jun 2012

20. EMSA. 2010. European Maritime Safety Agency. 2010. The 0,1% sulphur in fuel requirement as from 1 January 2015 in SECAs. An assessment of available impact studies and alternative means of compliance. Technical report. http://ec.europa.eu/environment/ air/transport/pdf/Report_Sulphur_Requirement.pdf . Accessed June 27, 2013

21. Endresen, Ø, E Sorgård, JK Sundet, SB Dahlsoren, ISA Isaksen, TF Berglen, and G Gravir. 2003. Emission from international sea transportation and environmental impact. Journal of Geophysical Research: Atmospheres 108 (D17):

22. Endresen, O, M Eide, S Dalsoren, and IS Isaksen. 2008. The environmental impacts of increased international maritime shipping. Past trends and future perspectives. http://www.oecd. org/env/transportandenvironment/41750201.pdf . Accessed 1 Nov 2012

23. European Commission. 2013. Clean power for transport. A European alternative fuels strategy. Communication from the Commission to the European parliament, the Council, the European Economic and Social Committee and the Committee of the regions. COM(2013) 17 final. http://eur-lex.europa.eu/ LexUriServ/LexUriServ.do?uri=COM:2013:0017:FIN:EN:PDF . Accessed Jul 1 2013

24. Eyring, V, HW Köhler, A Lauer, and B Lemper. 2005. Emissions from international shipping: 2. Impact of future technologies on scenarios until 2025. Journal of Geophysical Research: Atmospheres 110 (D17):

25. Forest Industries, Swedish Forest Industries Federation. 2012. Low sulphur fuel directive: EU industry competitiveness again disregarded. 25.5.2012. News published at: http://www. forestindustries.se/i_fokus_-_startsidenotiser_1/low-sulphur-fuel-directive-eu-industry-competitiveness-again-disregarded. Accessed 1 Nov 2012

26. Galaz, V, T Hahn, P Olson, C Folke, and U Svedin. 2008. The problem of fit among biophysical systems, environmental and resource regimes, and broader governance systems: insights and emerging challenges. In O. Young, LA King and H Schroder (eds). Institutions and environmental change: principal findings, applications, and research findings, pp. 147–186. Cambridge MA: MIT Press.

27. Gläser, J, and G Laudel. 2009. Experteninterviews und qualitative Inhaltsanalyse als Instrumente rekonstruierender Untersuchungen (3. Auflage) (Expert interviews and qualitative content analysis as instruments of reconstructive empirical investigations. Wiesbaden: VS Verlag für Sozialwissenschaften.

28. Gritsenko, D. 2013. The Russian dimension of Baltic maritime governance. Journal of Baltic Studies 44 (4): in press

29. Hajer, MA. 1995. The Politics of Environmental Discourse: Ecological Modernization and the Policy Process. Oxford: Oxford University Press.

30. Hall, P, and W Jacobs. 2010. Shifting proximities: the maritime ports sector in an era of global supply chains. Regional Studies 44 (9): 1103–1115.

31. Hassler, B. 2011. Accidental versus operational oil spills from shipping in the Baltic Sea: risk governance and management strategies. Ambio 40 (2): 170–178.

32. Hill, M, and DP Hupe. 2002. Implementing Public Policy: Governance in Theory and in Practice. London: SAGE.

33. Holma, E, and S Kajander. 2012. Baltic Port Barometer. Turku: University of Turku, Centre for Maritime Studies.

34. Izvestiya. 2012. "Sovkomflot" perevedet svoy Baltiyskiy flot na gas" (Sovcomflot will transfer its Baltic fleet to gas). Accessed 2 May 2013 http://izvestia.ru/news/541283

35. Jalkanen, JP, A Brink, J Kall, H Pettersson, J Kukkonen, and T Stipa. 2009. A modelling system for the exhaust emissions of maritime traffic and its application in the Baltic Sea area. Atmospheric Chemistry and Physics 9 (23): 9209–9223.

36. Jentoft, S. 2007. Limits of governability: institutional implications for fisheries and coastal governance. Marine Policy 31 (4): 360–370.

37. Kalli, J, T Karvonen, and T Makkonen. 2009. Sulphur content in ships bunker fuel in 2015. A study on the impacts of the new IMO regulations and transportation costs. Publications of the Ministry of Transport and Communications, 31. http://www.lvm.fi/c/document_library/get_file?folderId=339549&name=DLFE-8042.pdf&title=Julkaisuja%2031-2009 . Accessed 1 Jan 2012

38. Knutsen, OF, and B Hassler. 2011. IMO legislation and its implementation: accident risk, vessel deficiencies and national administrative practices. Marine Policy 35 (2): 201–207.

39. Kooiman, J. 1993. Modern Governance: New Government-Society Interactions. London: SAGE.

40. Kooiman, J. 2003. Governing as Governance. London: SAGE.

41. Kooiman, J, M Bavick, R Chuenpagdee, R Mahon, and R Pullin. 2008. Interactive governance and governability: an introduction. The Journal of Transdisciplinary Environmental Studies 7 (1): 1–11.

42. Lai, KH, V Lun, CWY Wong, and TCE Cheng. 2011. Green shipping practices in the shipping industry: Conceptualization, adoption, and implications. Resources, Conservation and Recycling 55 (6): 631–638.

43. Lascoumes, P, and P Le Gales. 2007. Introduction: Understanding Public Policy through Its Instruments—From the Nature of Instruments to the Sociology of Public Policy Instrumentation. Governance 20 (1): 1–21.

44. Lloyd's Register. 2011. Shipping and the environment. Issue 2011/2.   http://www.lr.org/Images/CD1925_LR_Marine_SATE_2%282%29_web_tcm155-216980.pdf. Accessed 27 Jun 2013

45. Maersk. 2011. Maesk Line Zero SOx programme prompts new fuel switch solution. News on the eyefortransport.com www-site, published March 8, 2011. http://www.eft.com/freight-transport/maersk%E2%80%99s-zero-sox-program-prompts-new-fuel-switch-solution. Accessed 27 Jun 2013

46. Maritime, UK. 2012. Sulphur regulations: bad for jobs... and the environment. http://www.maritimeuk.org/wp-content/uploads/2012/10/Maritime_UKsulphur.pdf. Accessed 1 Nov 2012

47. Marsden, G, and T Rye. 2010. The governance of transport and climate change. Journal of Transport Geography 18 (6): 669–678.

48. Miles, M, and A Huberman. 1994. Qualitative Data Analysis: An Expanded Sourcebook. London: SAGE.

49. Morskaya, B. 2011. Sokrashchaya vybrosy s morskix sudov (Reducing the emissions from maritime vessels). 2 (36): Accessed 2 May 2013 http://www.maritimemarket.ru/article.phtml?id=1351

50. Morskoy, B. 2012a. Ostrota ogranicheniy (The Sharpness of Limitation). Accessed 2 May 2013 http://www.mbsz.ru/28/57583.php

51. Morskoy, B. 2012b. SPG – toplivo budushchego? (LNG – The Fuel of the Future?). Accessed 2 May 2013 http://www.mbsz.ru/28/57570.php

52. Morskoy, F. 2012c. SPG v kachestve bunkernogo topliva – effektivnoe reshenie budushchego. (LNG as bunker fuel – effective solution for the future, vol. 5. Accessed 2 May 2013. http://www.morvesti.ru/analytics/index.php?ELEMENT_ID=17794

53. Morskoy, F. 2012d. Rynok bunkernogo topliva v Rossii: sostoyanie i perspektivy. (The bunker fuel market in Russia: Condition and perspectives), vol. 5. Accessed 2 May 2013. http://www.morvesti.ru/analytics/index.php?ELEMENT_ID=17686

54. Morskoy, F. 2012e. Baltika perexodit na SPG (Baltic Sea transfers to LNG), vol. 5. Accessed 2 May 2013. http://www.morvesti.ru/analytics/index.php?ELEMENT_ID=17883

55. Morskoy, F. 2012f. Pered bunkernym rynkom zamayachila problema 2015 (The problem 2015 arose in front of the bunker market), vol. 5. Accessed 2 May 2013. http://www.morvesti.ru/interview/index.php?ELEMENT_ID=17940

56. Moss, T. 2004. The governance of land use in river basins: prospects for overcoming problems of institutional interplay with the EU Water Framework Directive. Land Use Policy 21 (1): 85–94.

57. Moss, T. 2012. Spatial fit, from panacea to practice: implementing the EU water framework directive. Ecology and Society 17 (3): 2.

58. Ng, AKY, P Hall, and AA Pallis. 2013. Guest editors' introduction: institutions and the transformation of transport nodes. Journal of Transport Geography 27: 1–3.

59. North, D. 1990. Institutions. Cambridge: Cambridge University Press, Institutional Change and Economic Performance.

60. Oldakowski, B. 2012. LNG in the Baltic Sea project. Project overview. Warsaw, Poland: Presentation given at the Ministry of Transport, Constructions and Maritime Economy. http://www.transport.gov.pl/files/0/1795011/BogdanOldakowski02February2012Final.pdf . Accessed 6 Jun 2012

61. Olivier, D, and B Slack. 2006. Rethinking the port. Environment and Planning A 38 (8): 1409–1427.

62. Ostrom, E. 2010. Beyond markets and states: polycentric governance of complex economic systems. The American Economic Review 100 (3): 641–672.

63. Ostrom, E. 2011. Background on the institutional analysis and development framework. Policy Studies Journal 39 (1): 7–27.

64. Ostrom, E. 2012. Nested externalities and polycentric institutions: must we wait for global solutions to climate change before taking actions at other scales? Economic Theory 49 (2): 353–369.

65. Ostrom, V, CM Tiebout, and R Warren. 1961. The organization of government in metropolitan areas: a theoretical inquiry. American Political Science Review 55 (4): 831–842.

66. Pallis, AA. 2007. EU Port Policy Developments: Implications for Port Governance. In Issues on Devolution, Port Governance and Port Performance, Research in Transport Economics, vol 17, ed. Brooks MR, Cullinane K, 161–176. London: Elsevier.

67. PENTA. 2012. Project workshop "Major shifts or business as usual? Implications of sulphur regulation on maritime transport" Muuga, Estonia 18.4.2012. http://www.pentaproject.info/blog/workshop-in-tallinn-18-dot-4-presentations . Accessed 30 May 2013

68. Pike, K, N Butt, D Johnson, and S Walmsley. 2011. Global sustainable shipping initiatives. Audit and overview 2011. A Report for WWF. http://awsassets.panda.org/downloads/sustainable_shipping_initiatives_report_1.pdf . Accessed 27 Jun 2013

69. Ponte, S, and P Gibbon. 2006. Quality standards, conventions and the governance of global value chains. Economy and Society 34 (1): 1–31.

70. PortNews. 2013. Vysokie celi "nizkoy" sery (High Goals of "Low" Sulphur). Accessed 2 May 2013. http://portnews.ru/comments/1572/

71. PortStrategy. 2011. The EU ports vulnerable to sulphur regulation. http://www.portstrategy.com/news101/europe/sulphur-legislation-will-impact-ports . Accessed 6 Jun 2012

72. Rhodes, RAW. 1996. The New governance: governing without government. Political Studies 44 (4): 652–667.

73. Roe, M. 2008. Safety, security, the environment and shipping: the problem of making effective policies. WMU Journal of Maritime Affairs 7 (1): 263–279.

74. Roe, M. 2009. Multi-level and polycentric governance: effective policymaking for shipping. Maritime policy and management 36 (1): 39–56.

75. Roe, M. 2013. Maritime governance and policy-making. London: Springer Verlag.

76. Scharpf, FW. 1997. Games real actors play: Actor-centered institutionalism in policy research. Boulder: Westview Press.

77. Sheppard, E. 2002. The spaces and times of globalization: place, scale, networks and positionality. Economic Geography 78 (3): 307–330.

78. Skovgaard, J. 2012. Corporate social reposponsibility in the Danish shipping industry. Cambridge UK: Paper presented at DRUID Academy Conference 2012. http://druid8.sit.aau.dk/acc_papers/436055lixpr008mgnu655pfb6jhu.pdf. Accessed 27 Jun 2013

79. Small, ML. 2011. How to conduct a mixed methods study: recent trends in a rapidly growing literature. Annual Review of Sociology 37: 57–86.

80. SRO RAMRB. 2012. Samoreguliruemaya organizaciya Associaciya rossiyskix morskix I rechnyx bunkerovshchikov (Self-regulated organization Russian Association of Marine and River Bunker). : Open letter to Prime-Minister of the Russian Federation Dmitry Medvedev. Accessed 2 May 2013. http://mrbunker.ru/Image/%D0%B8%D1%81%D1%85_%20218.pdf

81. Stankiewicz, M. 2012. HELCOM perspective on clean Baltic Sea shipping. Riga Latvia: Presentation given at CLEANSHIP Mid-Term Conference. http://www.clean-baltic-sea-shipping.com/uploads/files/Monika_Stankiewicz_Clean_Shipping_mid_conference_2012.pdf. Accessed 24 Oct 2012

82. Stenhede, T. 2012. New fuels for ship engines. Riga Latvia: Presentation given at CLEANSHIP Mid-Term Conference. http://www.clean-baltic-sea-shipping.com/uploads/files/SPIRETH_Riga_seminar_2_pptx.pdf. Accessed 24 Oct 2012

83. Stubbe, W. 2012. Cross-border cooperation of small & medium-sized ports. Baltic Transport Journal 48 (4): 22.

84. Sweco. 2012. Consequences of the sulphur directive. http://www.svensktnaringsliv.se/multimedia/archive/00033/Consequences_of_the__33781a.pdf Accessed 30 Oct 2012

85. Tan, AKJ. 2006. Vessels-Source Marine Pollution. The Law and Politics of International regulation. Cambridge: Cambridge University Press.

86. TEN-T EA, Trans-European Transport Network Executive Agency. 2012. LNG in Baltic Sea Ports. http://tentea.ec.europa.eu/en/ten-t_projects/ten-t_projects_by_country/multi_country/2011-eu-21005-s.htm. Accessed 24 Oct 2012

87. The Ports of Stockholm. 2011. Joint Baltic Sea collaboration between Ports of Stockholm and the Port of Turku. http://www.stockholmshamnar.se/en/News-and-press/2011/Joint-Baltic-Sea-collaboration-between-Ports-of-Stockholm-and-the-Port-of-Turku. Accessed 6 Jun 2012

88. Torfing, J, BG Peters, J Pierre, and E Sorensen. 2012. Interactive governance. Advancing the paradigm. Oxford: Oxford University Press.

89. Transport & Environment. Newsroom. 2012. EU agrees to significant sulphur reduction in shipping fuels. http://www.

transportenvironment.org/news/eu-agrees-significant-sulphur-reduction-shipping-fuels. Accessed 6 Jun 2012

90. Verhoeven, P. 2010. A review of port authority functions: towards a renaissance? Maritime Policy & Management 37 (3): 247–270.

91. Viking Line. 2013a. Viking Line an environmental pioneer with its LNG vessel. Information at company www-page. http://www.vikingline.com/en/Investors-and-the-Group/Safety--environment/Environment/Viking-Grace/. Accessed 27 Jun 2013

92. Viking Line. 2013b. Press release April 18, 2013. M/S Viking Grace increased Viking Line's market share. http://www.vikingline.com/Documents/pressreleases/20130418-viking-grace-marknadsandel-en.pdf. Accessed 27 Jun 2013

93. Wahlström, J, N Karvosenoja, and P Porvari. 2006. Ship emissions and technical emission reduction potential in the Northern Baltic Sea. 8. Reports of Finnish Environment Institute 2006. http://cleantech.cnss.no/wp-content/uploads/2011/06/2006-Wahlstrom-ships-emissions-technical-emission-reduction-potential-in-northern-Baltic-Sea.pdf. Accessed 26 Jun 2013

94. Wallenius, W. 2013. Environmental objectives. http://www.2wglobal.com/www/environment/objectives/index.jsp. Accessed 27 Jun 2013

95. Wärtsilä. 2010. Exhaust gas scrubber installed onboard MT "Suula". Public test report. http://www.wartsila.com/en/search?q=Exhaust+gas+scrubber+installed+onboard+MT+Suula. Accessed 27 Jun 2013

96. Weber, R, P Galera-Lindblom, and RO Rasmussen. 2012. Green growth in the Baltic Sea region. State of the Region Report 2012. The top of Europe bracing itself for difficult times: Baltic Sea region –collaboration to sustain growth. Nordic Investment Bank, European Investment Bank and Baltic Development Forum.

97. Wheeler, MF, and M Peszynska. 2002. Computational engineering and science methodologies for modeling and simulation of subsurface applications. Advances in Water Resources 25 (8–12): 1147–1173.

98. Wuisan, L, J Van Leuwen J, and CSA Koppen. 2012. Greening international shipping through private governance: a case study of the CleanShipping Project. Marine Policy 36 (1): 165–173.

99. Yle Uutiset. 2012. EU parliament approves sulphur directive. http://yle.fi/uutiset/eu_parliament_approves_sulphur_ directive/6290175. Accessed 17 Sep 2012

100. Young, OR. 1994. International Governance: Protecting the Environment in a Stateless Society. NY: Cornwell University Press.

Corrosion Inhibitive Properties of Some New Isatin Derivatives on Corrosion of N80 Steel in 15% HCl

Mahendra Yadav[1], Usha Sharma[1], and Premanand Yadav[2]

[1]Department of Applied Chemistry, Indian School of Mines, Dhanbad 826004, India
[2]Department of Physics, Post Graduate College Ghazipur, Ghazipur 233002, India

ABSTRACT

Background

The inhibition effect of two synthesized isatin compounds, namely 1-morpholinomethyl-3(1-N-dithiooxamide)iminoisatin [MMTOI] and

1-diphenylaminomethyl-3(1-*N*-dithiooxamide)iminoisatin [PAMTOI], on the corrosion inhibition of N80 steel in 15% HCl solution was studied by polarization, alternating current impedance (electrochemical impedance spectroscopy), and weight loss measurements. The surface examination was carried out by scanning electron microscopy and Fourier transform infrared spectroscopy.

Results

The compounds [PAMTOI] and [MMTOI] show the maximum of 91.2% and 84.3% inhibition efficiency, respectively, at 200-ppm concentration. Polarization curves revealed that the used inhibitors represent mixed-type inhibitors. Adsorption of used inhibitors led to a reduction in the double-layer capacitance and an increase in the charge transfer resistance.

Conclusions

Results show that both inhibitors were effective inhibitors and their inhibition efficiency was significantly increased with increasing concentration. Adsorption of both compounds obeys the Langmuir adsorption isotherm.

BACKGROUND

N80 steel is widely used as a construction material for pipe work in the oil and gas production, such as downhole tubular, flow lines, and transmission pipelines in the petroleum industry. Mineral acids, particularly hydrochloric acid, are frequently used in industrial processes involving acid cleaning, acid pickling, acid descaling, and oil well acidizing [1-3]. In the petroleum industry, 15% HCl is commonly used for acidizing treatment because it leaves no insoluble product after the treatment and is found to be commercially available and cheap, but adversely at the same time, it severely attacks the metal casings and tubular of oil well during the acidizing process. Therefore, protective measures should be required to prevent the metal loss due to corrosion by using chemical and other means. Due to the aggressiveness of acids, inhibitors are often used to reduce the rate of dissolution of metals. Most of the well-known acid inhibitors are organic compounds containing nitrogen, oxygen, and/or sulfur atoms; heterocyclic compounds; and delocalized ϖ-electrons [4-12]. The polar functional groups are usually regarded as the reaction center for the establishment of the adsorption process [13]. It is generally accepted that organic molecules inhibit corrosion via adsorption at the metal-solution interface [14,15], making the adsorption layer to function as a barrier and isolating the metal from the corrosion [16]. Some Mannich bases have been reported as efficient corrosion inhibitors [17,18], and the literature available to date about the Mannich bases used as corrosion inhibitors is limited. Keeping in view the importance of Mannich bases as potential corrosion inhibitors, we have synthesized k;1-morpholinomethyl-3(1-N-dithiooxamide)iminoisatin   [MMTOI] and   1-diphenylaminomethyl-3(1-N-dithiooxamide)iminoisatin [PAMTOI] and studied their corrosion inhibition properties for N80 steel in 15% HCl solution.

METHODS

Materials and Surface Preparation

The working electrode for electrochemical studies and specimens for weight loss experiments were prepared from oil-well N80 steel sheets (supplied by ONGC, Dehradun, India) having the following percentage by weight (wt.%) composition: C, 0.31; Mn, 0.92; Si, 0.19; P, 0.01; S, 0.008; Cr, 0.20; and Fe, remainder. The specimens were mechanically polished with different grades of silicon carbide papers, degreased in ethanol to obtain a fresh oxide-free surface, washed with double-distilled water, and dried at room temperature.

Weight loss measurements

The specimens for the weight loss measurements were of the size 3 cm × 3 cm × 0.1 cm. Both sides of the specimens were exposed for weight loss measurement. For weight loss experiments, 300 mL of 15% HCl (v/v) was taken in 500-mL glass beakers. The inhibition efficiencies ( )

were evaluated after a pre-optimized time interval of 6 h using 20, 50, 100, 150, and 200 ppm of inhibitors. The specimens were removed from the electrolyte, washed thoroughly with distilled water, dried, and weighed. The inhibition efficiencies were evaluated using the following formula:

$$(\%)=[(W-W_i)/W]\times100 \tag{1}$$

Where W is the weight loss in the absence of an inhibitor and Wi is the weight loss in the presence of an inhibitor.

The corrosion rate ( ) of the specimen can be calculated with the help of the following equation:

$$(mpy)= \frac{534W}{DAT}, \tag{2}$$

Where W is the weight loss (mg), D is the density of the specimen (g cm^{-3}), A is the area of the specimen (cm^2), and T is the exposure time (h).

Electrochemical Studies

The electrochemical experiments were carried out in a three-necked glass assembly containing 150 mL of the electrolyte with different concentrations of inhibitors (from 20 to 200 ppm by weight) dissolved in it. The potentiodynamic polarization studies were carried out with N80 steel working electrodes having an exposed area of 1 cm^2. A conventional three-electrode cell consisting of N80 steel as working electrode, platinum as counter electrode, and a saturated calomel electrode as reference electrode was used. Polarization studies were carried out using a VoltaLab 10 electrochemical analyzer (Radiometer Analytical, Lyon, France), and data were analyzed using the Voltamaster 4.0 software (Radiometer Analytical). The potential sweep rate was 10 mV s^{-1}. All experiments were performed at 298 K in an electronically controlled air thermostat. For calculating inhibition efficiency by the electrochemical polarization method, the following formula was used:

$$(\%)=[(I_0-I_{inh})]/I_0\times100 \qquad (3)$$

Where I_0 is the corrosion current density in the absence of an inhibitor and I_{inh} is the corrosion current density in the presence of an inhibitor.

AC Impedance Studies

Alternating current (AC) impedance studies were carried out using the same instrument as mentioned in polarization studies. The electrochemical impedance spectra (EIS) were acquired in the frequency range of 10 kHz to 1 mHz at the rest potential by applying 5-mV sine-wave AC voltage. The charge transfer resistance (R_{ct}) and double-layer capacitance (C_{dl}) were determined from Nyquist plots. The inhibition efficiencies were calculated from R_{ct} values using the following formula:

$$(\%)=[(R_{ct(inh)}-R_{ct})/R_{ct(inh)}]\times100 \qquad (4)$$

Where R_{ct} is the charge transfer resistance in the absence of an inhibitor and $R_{ct(inh)}$ is the charge transfer resistance in the presence of an inhibitor.

Synthesis of Inhibitors

The isatin Mannich bases, namely 1-morpholinomethyl-3(1-N-dithiooxamide)iminoisatin [MMTOI] and 1-diphenylaminomethyl-3(1-N-dithiooxamide)iminoisatin [PAMTOI], were synthesized by the reported method [19]. Isatin and dithiooxamide in 1:1 molar ratio were refluxed in ethanol for 8 h and cooled, and the precipitate was filtered. This product was subsequently treated with formaldehyde and morpholine or diphenylamine to get the desired product. The compounds were characterized by Fourier transform infrared (FTIR) spectroscopy, and purity of the compounds was checked by TLC. The names and molecular structures of the studied compounds are given in Figure 1.

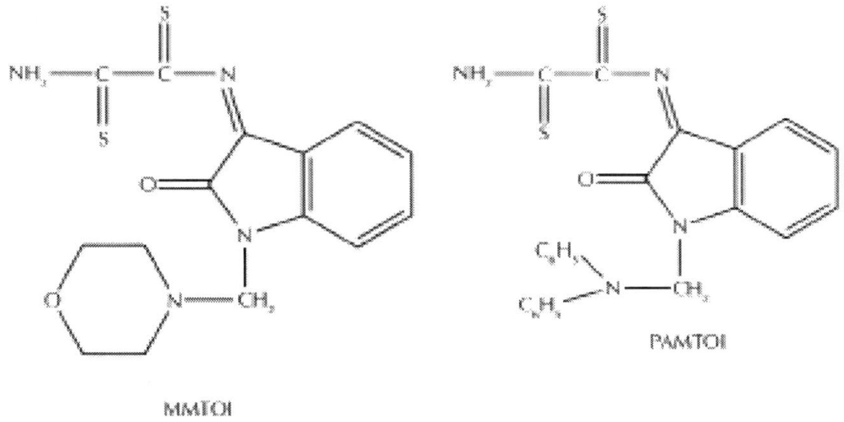

Figure 1: Molecular structures of [MMTOI] and [PAMTOI].

Scanning Electron Microscopy

The surface examination was carried out using a scanning electron microscope (JEOL 5400, Akishima-shi, Japan); the energy of the acceleration beam employed was 30 kV. All micrographs of the corroded specimens were carried out at a magnification of ×1,000.

FTIR Study

The N80 steel specimen was immersed in 15% HCl solution containing optimum concentration of inhibitors for 6 h. The specimen was taken out and dried, and the film was scraped using a non-metallic scrapper. FTIR spectra for the pure sample and the scraped films from the inhibited specimen were recorded using a PerkinElmer FTIR (Spectrum 2000) by KBr pellet method.

RESULTS AND DISCUSSION

Weight Loss Study

Weight loss studies were performed in accordance with the ASTM method. Tests were conducted in 15% HCl (*v/v*) solution for 6 h of exposure time at different concentrations of inhibitors (20 to 200 ppm) and temperatures (298 to 333 K).

Effect of Concentration

Both inhibitors were tested for 6 h of exposure period at different concentrations, and their corresponding weight loss data are presented in Table 1. The compounds [PAMTOI] and [MMTOI] have the maximum of 91.2% and 84.3% inhibition efficiency, respectively, at 200-ppm concentration. The inhibition efficiency of both inhibitors increases with increasing concentration of inhibitors (Figure 2), indicating that adsorption of inhibitors increases as concentration increases, resulting in the reduction of corrosion rate.

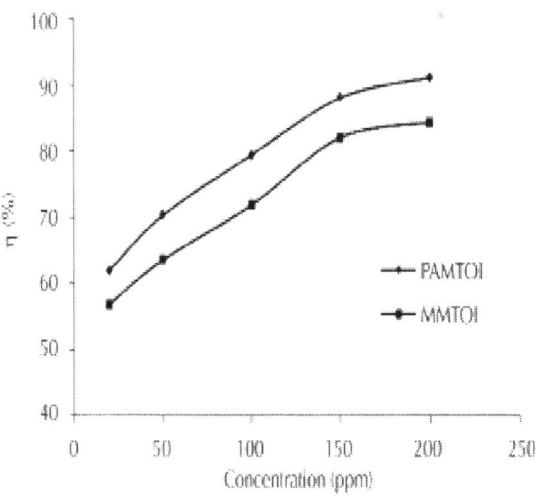

Figure 2: Variation of inhibition efficiency with concentration in the presence of [PAMTOI] and [MMTOI].

Table 1: Corrosion inhibition of N80 steel in 15% HCl in the absence and presence of inhibitors

Concentration	[PAMTOI]		[MMTOI]	
(ppm)	(mpy)	(%)	(mpy)	(%)
0	9.55	-	9.55	-
20	3.62	62.1	4.11	56.8
50	2.83	70.3	3.47	63.6
100	1.96	79.4	2.69	71.8
150	1.13	88.1	1.72	82.0
200	0.84	91.2	1.49	84.3

Yadav *et al.*

Yadav *et al.* *International Journal of Industrial Chemistry* 2013 4:6, doi:10.1186/2228-5547-4-6

Effect of Temperature

The effect of temperature on the corrosion inhibition efficiency of [PAMTOI] and [MMTOI] for N80 steel in 15% HCl was investigated by weight loss measurements in the temperature range of 298 to 333 K in the absence and presence of both inhibitors at optimum concentration (200 ppm) in a thermostat. Table 2 shows values of corrosion rate ( ) and inhibition efficiency ( ) at different temperatures for both inhibitors. The observation depicts that the rate of corrosion increases with increase in temperature. At this temperature, the metal organic complex layer dissociates leaving a porous diffused film, which is responsible for corrosion [20].

Table 2: Corrosion parameters in the presence and absence of [PAMTOI] and [MMTOI] at different temperatures

Temperature	Blank	[PAMTOI]		[MMTOI]	
(K)	ρ (mpy)	ρ (mpy)	η (%)	ρ (mpy)	η (%)
298	9.55	0.84	91.2	1.49	84.3
303	12.09	1.37	88.6	2.28	81.1
313	19.27	2.95	84.7	4.46	76.9

| 323 | 30.42 | 6.13 | 79.9 | 8.41 | 72.3 |
| 333 | 48.94 | 11.57 | 76.4 | 15.22 | 68.9 |

Yadav et al.

Yadav et al. International Journal of Industrial Chemistry 2013 4:6, doi:10.1186/2228-5547-4-6

Adsorption Isotherms

The adsorption isotherm experiments were performed to have more insights into the mechanism of corrosion inhibition since it describes the molecular interaction of the inhibitor molecule with the active sites on the N80 steel surface [21]. To determine the adsorption mode, various isotherms were tested. Langmuir adsorption isotherms were found to be the best, which give a straight line graph for the plot of log (/1 −) versus the logarithmic concentration of inhibitors (Figure 3). The inhibition of the corrosion of metals by organic inhibitors is usually attributed to either the adsorption of the inhibitor molecules or the formation of a barrier layer of insoluble metal complexes.

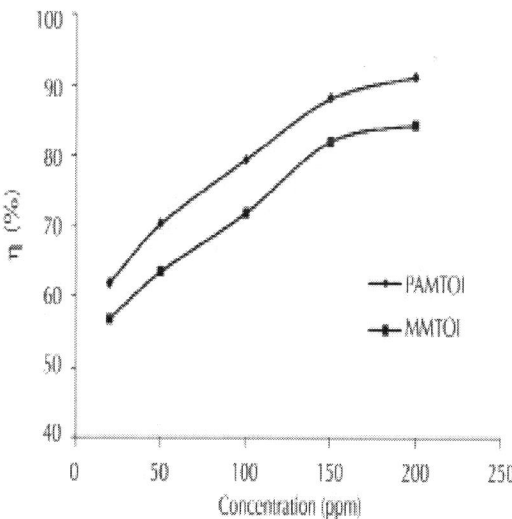

Figure 3: Langmuir adsorption isotherms in the presence of [PAMTOI] and [MMTOI].

It is well recognized that organic inhibitor molecules set up their inhibition action via the adsorption of the inhibitor molecules onto the metal/solution interface. The adsorption process is affected by the chemical structures of the inhibitors, the nature and charged surface of the metal, and the distribution of charge over the whole inhibitor molecule. In general, owing to the complex nature of adsorption and inhibition of a given inhibitor, a single adsorption mode between inhibitor and metal surface is impossible. Organic inhibitor molecules may be adsorbed on the metal surface in one or more ways: (a) electrostatic interaction between the charged molecules and the charged metal, (b) interaction of unshared electron pairs in the molecule with the metal, (c) interaction of ϖ-electrons with the metal, or (d) a combination of types (a) to (c) [22,23]. In the aqueous acidic solutions, [MMTOI] and [PAMTOI] exist either as neutral molecules or as protonated molecules (cations). Generally, two modes of adsorption could be considered. In one mode, the neutral inhibitors may be adsorbed on the surface of N80 steel through the chemisorption mechanism, involving the displacement of water molecules from the N80 steel surface and the sharing of electrons between the heteroatoms and iron. The inhibitor molecules can also adsorb on the N80 steel surface on the basis of donor-acceptor interactions between ϖ-electrons of the aromatic ring and vacant d orbital of the surface iron. In another mode, since it is well known that the steel surface bears a positive charge in acid solution [24], it is difficult for the protonated inhibitors to approach the positively charged mild steel surface (H_3O^+/metal interface) due to electrostatic repulsion. Since chloride ions have a smaller degree of hydration, they bring excess negative charges in the vicinity of the interface and favor more adsorption of the positively charged inhibitor molecules; the protonated inhibitors adsorb through electrostatic interactions between the positively charged inhibitor molecules and the negatively charged metal surface. Thus, there is a synergism between adsorbed Cl^- ions and protonated inhibitors. Experimental data reveal that the inhibition efficiency of inhibitors follows the order [PAMTOI] > [MMTOI]. This order of performance is best explained in terms of the size of the inhibitors. Both inhibitors have the same number of active centers, but the size of [PAMTOI] is larger than that of [MMTOI], so the inhibition efficiency of [PAMTOI] is greater than that of [MMTOI].

The inhibition efficiency afforded by [MMTOI] and [PAMTOI] may be attributed to the presence of electron-rich N atom and aromatic rings. One phenylimino group and one indoline ring are common in the structure of both inhibitors. Therefore, the possible reaction centers are unshared electron pairs of sulfur of the C=S group, nitrogen of the -NH$_2$ group and the C=N group, and ϖ-electrons of the aromatic ring.

Kinetic and Thermodynamic Study

The apparent activation energy (E_a) for the dissolution of N80 steel in 15% HCl was calculated from the slope of plots using the Arrhenius equation:

$$\text{Log } k = -E_a/2.303RT + \log A \qquad (5)$$

Where k is the rate of corrosion, E_a is the apparent activation energy, R is the universal gas constant, T is the absolute temperature, and A is the Arrhenius pre-exponential factor. By plotting log k against $1/T$, the values of E_a have been calculated (E_a = $-$(Slope) × 2.303 × R; Figure 4). The activation energy for the reaction of N80 steel in 15% HCl increases in the presence of inhibitors. The E_a values for [PAMTOI] and [MMTOI] were found to be higher than the E_a value for the blank solution (Table 3). The values of E_a for [PAMTOI] and [MMTOI] were found to be 61.46 and 50.03 kJ mol^{-1}, respectively.

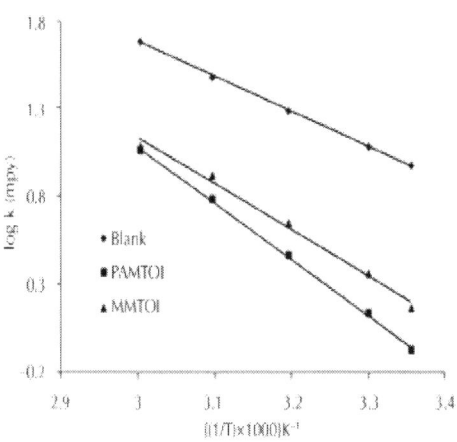

Figure 4: Arrhenius plot for corrosion of N80 steel in 15% HCl in the presence/absence of inhibitors.

Table 3: Thermodynamic parameters in the absence and presence of [PAM-TOI] and [MMTOI]

Inhibitor	H* (kJ mol−1)	Ea(kJ mol−1)	Gads (kJ mol−1)	S * (J mol−1 K−1)
Blank	-	38.38	-	-
[PAMTOI]	−58.74	61.46	−38.15	−49.42
[MMTOI]	−51.64	50.03	−34.11	−68.62

Yadav et al.

Yadav et al. International Journal of Industrial Chemistry 2013 4:6, doi:10.1186/2228-5547-4-6

The val4ues of the entropy of activation (ΔS^*) and enthalpy of activation (ΔH^*) were calculated using the following formula:

$$k= (RT/Nh)\ exp\ (\ S^*/R)\ exp\ (-\ H^*/RT) \tag{6}$$

Where k is the rate of corrosion, h is Planck's constant, N is Avogadro's number, S^* is the entropy of activation, and H^* is the enthalpy of activation. A plot of log (k/T) versus $1/T$ (Figure 5) gives a straight line, with a slope of ($-\Delta H^*/2.303R$) and an intercept of [log ($R/N\ h$) + $S^*/2.303R$], from which the values of S^* and H^* were calculated (Table 3). The H^* values for [PAMTOI] and [MMTOI] were found to be −58.74 and −51.64 kJ mol^{-1}, respectively, which revealed the exothermic nature of the corrosion reaction. The S^* values were found to be −49.42 and −68.62 J mol^{-1} K^{-1}for [PAMTOI] and [MMTOI], respectively (Table 3). The negative values of S^* for [PAMTOI] and [MMTOI] imply that the activated complex in the rate-determining step represents an association rather than a dissociation step, indicating that a decrease in disordering takes place on going from reactants to the activated complex [25,26]. This also indicated that the activated complex was in a higher order state than at the initial state.

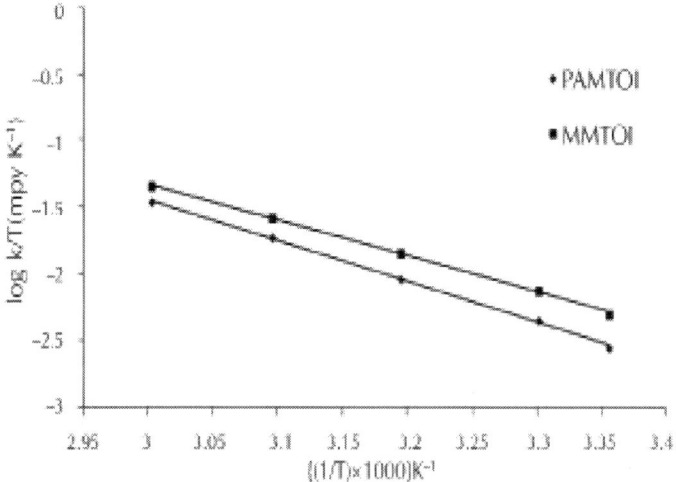

Figure 5: A plot of log k/T versus $1/T$ in the presence of [PAMTOI] and [MMTOI].

The average values for the change in the free energy of adsorption (G_{ads}) were calculated using the following equation:

$$K = \frac{1}{55.5} \exp \left(\frac{-\Delta G_{ads}}{RT} \right),$$

(7)

Where R is the molar gas constant (J K^{-1} mol^{-1}) and T is the temperature. The value 55.5 in the above equation is the concentration of water in the solution (mol L^{-1}). The equilibrium constant (K) has been replaced by the following equation:

$$K = / (1 -) C,$$

(8)

Where is the degree of coverage on the metal surface and C is the concentration of inhibitors.

By plotting log K against $1/T$, the value of G_{ads} was calculated ($G_{ads} = -2.303 \times R \times$ Slope) from the slope of the straight line obtained (Figure 6). The values of the free energy of adsorption at 200 ppm of [PAMTOI] and [MMTOI] were found to be −38.15 and −34.11 kJ mol^{-1}, respectively. The negative values of G_{ads} are consistent with the spontaneity of the adsorption process on the N80 steel surface. Generally, the change in free energy values of −20 kJ mol^{-1} or less negative are associated with an electrostatic interaction between charged molecules and charged metal surface (physisorption); those above −40 kJ mol^{-1} or more negative involve charge sharing or transfer from the inhibitor molecules to the metal surface to form a coordinate covalent bond (chemisorption) [27,28]. In the present study, the G_{ads} values obtained for both inhibitors on N80 steel in 15% HCl solution were higher than −20 kJ mol^{-1} but less than −40 kJ mol^{-1} (Table 3); this indicates that the adsorption is neither typical physisorption nor typical chemisorption, but it is of a complex mixed type, that is the adsorption of inhibitor molecules on the N80 steel surface in the present study involves both physisorption and chemisorption. Lebrini et al. [29] studied some triazole derivatives as corrosion inhibitors of mild steel in 1 M HClO$_4$. The Gibbs free energy of adsorption of these molecules was reported to be around −34 kJ mol^{-1}. They concluded that the adsorption mechanism of these molecules on steel involved two types of interactions, chemisorption and physisorption. A similar conclusion was found by Ozcan [30], who studied the use of cystine as a corrosion inhibitor on mild steel in sulfuric acid. Thus, adsorption of the studied inhibitors at the surface of N80 steel is not pure physisorption, but it is a combination of physisorption as well as chemisorption.

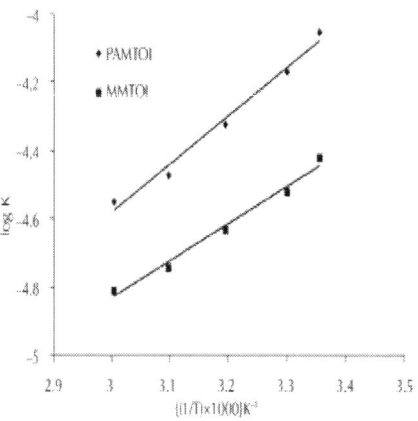

Figure 6: A plot of log K_{equ} versus $1/T$ in the presence of [PAMTOI] and [MMTOI].

Potentiodynamic Polarization Study

Figures 7 and 8 show the polarization curves of N80 steel in 15% HCl solution in the absence and presence of different concentrations (20, 100, and 200 ppm) of [PAMTOI] and [MMTOI], respectively. The polarization curves remain almost the same in the absence and presence of both inhibitors, but in the presence of inhibitors, the curves shifted towards the lower current density as compared to the blank solution. The shift in current density to a lower value is high with increasing concentration of the inhibitors. The electrochemical corrosion parameters from the polarization curves of [PAMTOI] and [MMTOI] are presented in Table 4. It is apparent from Table 4 that I_{corr} decreases considerably in the presence of both inhibitors and increases with the increase in the inhibitor concentration due to the increase in the blocked fraction of the electrode surface by adsorption. The variation in the values of $_a$ and $_c$ in the presence of both inhibitors may indicate that both the anodic and cathodic processes are controlled. A minor shift in E_{corr} values towards the negative direction was obtained in the presence of both inhibitors, indicating the mixed nature of the inhibitors. Generally, if the displacement in E_{corr} is >85 mV with respect to E_{corr} in the uninhibited solution, the inhibitor can be seen as a cathodic or anodic type [31,32]. In our study, the maximum displacement is 22 mV in [PAMTOI] and 18 mV in [MMTOI], which indicates that both inhibitors can be arranged as a mixed-type inhibitor.

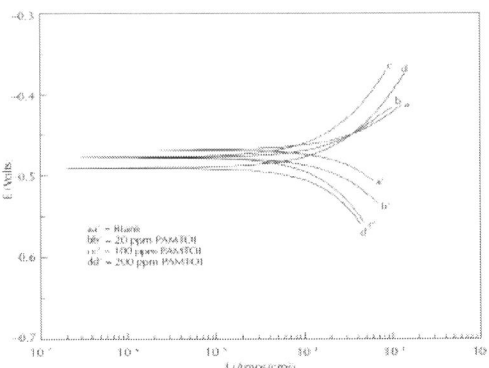

Figure 7: Potentiodynamic polarization curves for N80 steel in 15% HCl in the absence and presence of [PAMTOI].

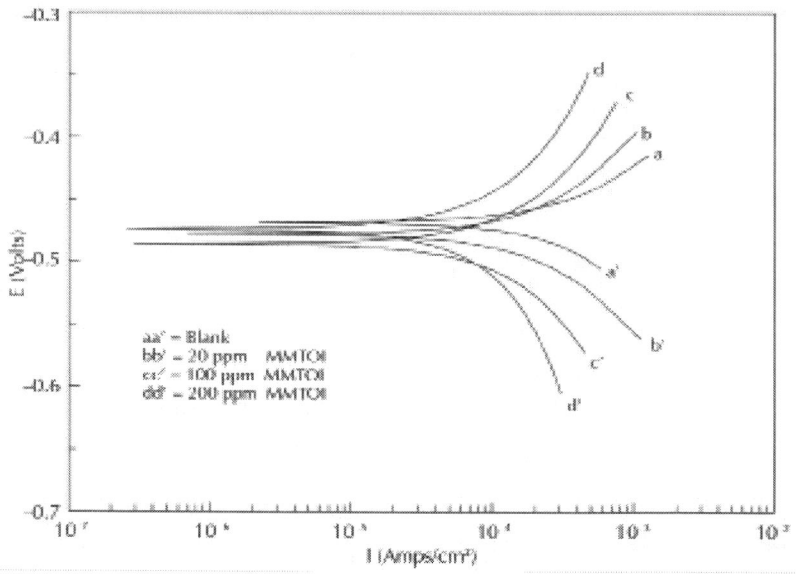

Figure 8: Potentiodynamic polarization curves for N80 steel in 15% HCl in the absence and presence of [MMTOI].

Table 4: Electrochemical corrosion parameters in the absence and presence of [PAMTOI] and [MMTOI]

Inhibitors	Concentration	Tafel slopes		Icorr	Ecorr		(%)
	(ppm)	a	c	(µA cm−2)	(mV)	(%)	wt. loss
		(mV dec−1)	(mV dec−1)				
Blank	-	109	153	471	−468	-	-
[PAMTOI]	20	112	167	187	−477	64.2	62.1
	100	116	179	88	−477	81.2	79.4
	200	125	186	36	−490	92.2	91.2
[MMTOI]	20	115	165	197	−477	58.2	56.8
	100	120	173	131	−486	72.1	71.8
	200	123	182	63	−473	86.5	84.3

Yadav et al.

Yadav et al. International Journal of Industrial Chemistry 2013 4:6, doi:10.1186/2228-5547-4-6

Electrochemical Impedance Spectroscopy

Nyquist plots of N80 steel in 15% HCl in the presence and absence of different concentrations (20, 100, and 200 ppm) of [PAMTOI] and [MMTOI] at 298 K are shown in Figures 9 and 10, respectively. All Nyquist plots obtained were semicircle in nature; the diameter of the semicircles increased with increase in inhibitor concentration, and the shape was maintained throughout the tested concentration, indicating that almost no change in the corrosion mechanism occurred due to inhibitor action. The values of electrochemical C_{dl} were calculated at the frequency f_{max}, at which the imaginary component of the impedance is maximal $(-Z_i)$ using the following equation:

$$C_{dl}=1/2\varpi f_{max}R_{ct} \tag{9}$$

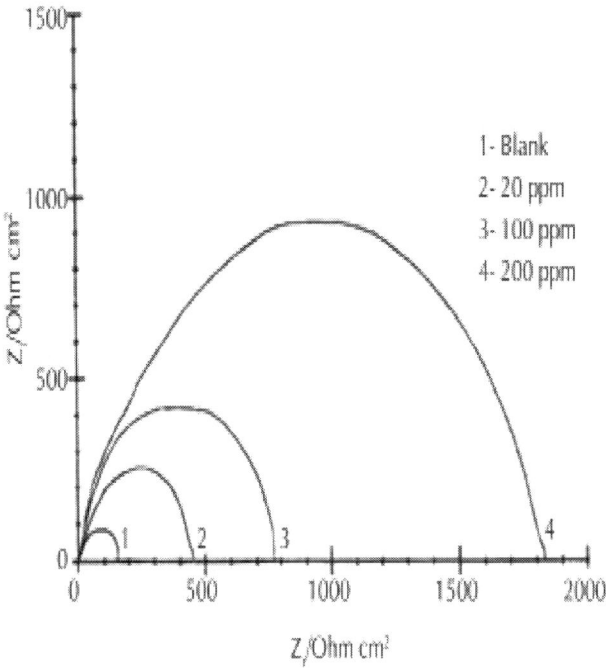

Figure 9: Nyquist plots of the corrosion of N80 steel in 15% HCl without and with [PAMTOI].

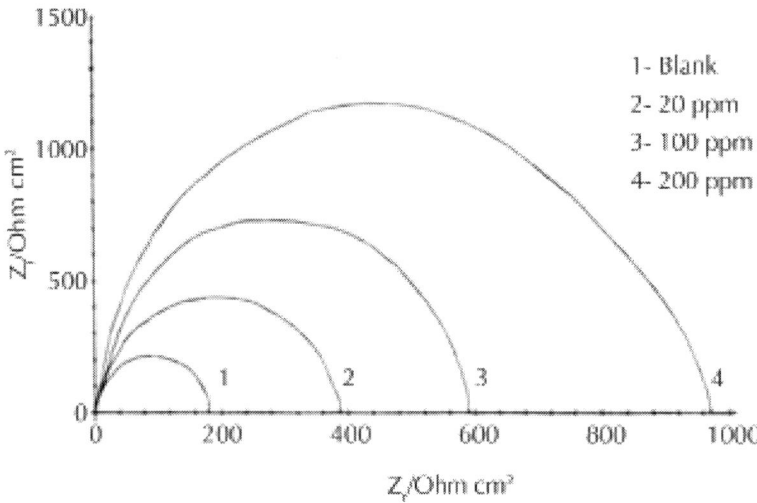

Figure 10: Nyquist plots of the corrosion of N80 steel in 15% HCl without and with [MMTOI].

The electrochemical parameters (R_{ct}, C_{dl}, and) calculated from the Nyquist plots using the equivalent circuit (Figure 11) in the presence and absence of inhibitors are presented in Table 5. In the EIS study, an increase in the R_{ct} value is observed with increasing inhibitor concentration, suggesting that the charge transfer process is retarded due to a decrease in the uncovered surface available for corrosion reaction. In this case, the higher the inhibitor concentration, the lower is the associated corrosion rate. Due to the non-homogeneity or roughness of the metal surface, the observed semicircles of capacitive loops were depressed into Z_i which is often referred to as frequency dispersion [33]. It is worth mentioning that the C_{dl} value is affected by imperfections of the surface. The C_{dl} values were found to decrease with increase in concentration of inhibitor solutions. This behavior is generally seen for a system where inhibition occurred due to the formation of a surface film by the adsorption of inhibitor on the metal surface [34]. The decrease in C_{dl}, which results from a decrease in the local dielectric constant and/or an increase in the thickness of the electrical double layer, suggests that the inhibitor molecules act by adsorption at the metal/solution interface [35]. Inhibition efficiencies obtained from weight loss, potentiodynamic polarization curves, and EIS were found to be in good agreement.

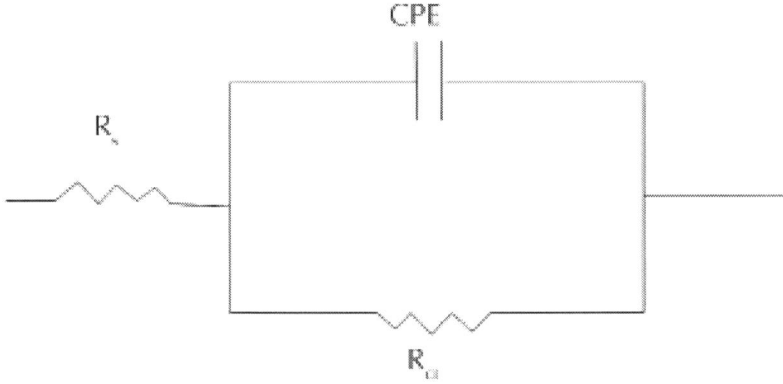

Figure 11: Equivalent circuit diagram.

Table 5: Electrochemical impedance parameters in the absence and presence of [PAMTOI] and [MMTOI] at different concentrations

Inhibitors	Concentration (ppm)	Rct (Ω cm2)	Cdl (μF cm-2)	(%)
Blank	-	176	662	-
[PAMTOI]	20	440	279	60.1
	100	780	158	77.5
	200	1825	79	90.3
[MMTOI]	20	390	320	54.9
	100	583	225	69.8
	200	980	190	82.1

Yadav *et al.*

Yadav *et al. International Journal of Industrial Chemistry* 2013 4:6, doi:10.1186/2228-5547-4-6

FTIR Analysis of Corrosion Products

In order to evaluate the protective layer formed on the steel surface in the presence of inhibitors and also to provide new bonding information on steel surface, a FTIR study was done. The FTIR spectra of pure and metal surface products after the corrosion test of [PAMTOI] and [MMTOI]

are shown in Figures 12 and 13, respectively. FTIR absorption band positions of the pure sample of [PAMTOI] with respect to that of its metal surface product (Figure 12) were observed to have shifted from 3,450 to 3,419 cm^{-1}, 2,925 to 2,924 cm^{-1}, 2,856 to 2,847 cm^{-1}, 1,744 to 1,723 cm^{-1}, 1,646 to 1,619 cm^{-1}, 1,458 to 1,455 cm^{-1}, 1,397 to 1,339 cm^{-1}, and 883 to 870 cm^{-1} corresponding to -NH, C-H aromatic, C-H methylene, C=O, C=N, C-N, C=S, and C-S stretching vibrations, respectively. FTIR absorption band positions of the pure sample of [MMTOI] with respect to that of its metal surface product (Figure 13) were observed to have shifted from 3,445 to 3,416 cm^{-1}, 2,925 to 2,920 cm^{-1}, 2,857 to 2,850 cm^{-1}, 1,741 to 1,729 cm^{-1}, 1,647 to 1,636 cm^{-1}, 1,560 to 1,550 cm^{-1}, 1,160 to 1,145 cm^{-1}, 1,487 to 1,465 cm^{-1}, and 835 to 802 cm^{-1} corresponding to -NH, C-H aromatic, C-H methylene, -C=O, -C=N, -C=C, -C-S, -C=S, and morpholine stretching vibrations, respectively [36]. The major shift in the position of bands corresponding to C=N and -NH indicates involvement of these groups in adsorption, and the minor shift in position of bands of the other groups indicates the interaction of the inhibitor molecule with the atoms of the metal surface.

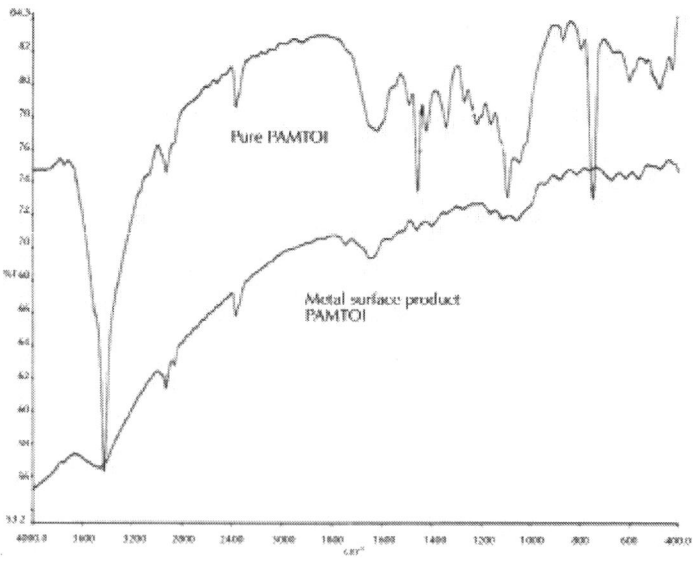

Figure 12: FTIR spectrum of pure and metal surface product obtained after corrosion in acid containing [PAMTOI].

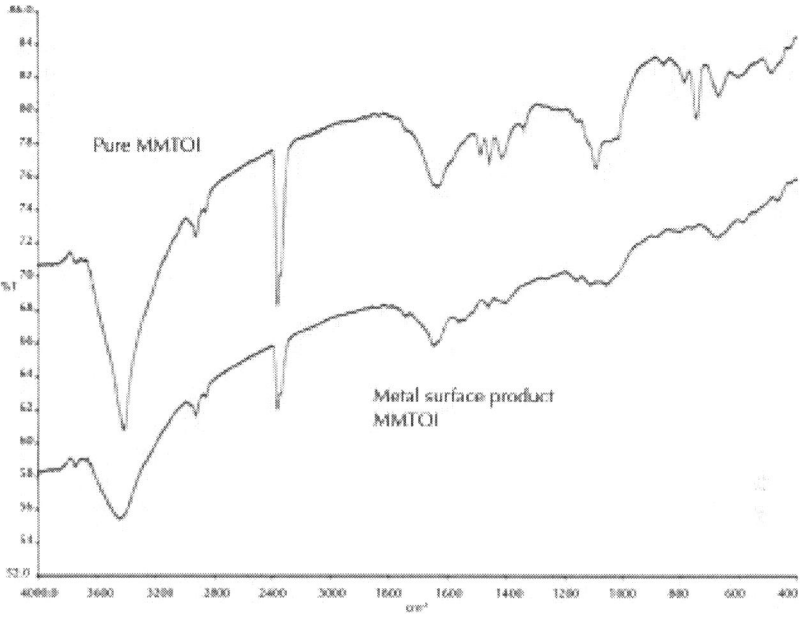

Figure 13: FTIR spectrum of pure and metal surface product obtained after corrosion in acid containing [MMTOI].

SEM Micrographs of Metal Surface

Figure 14a shows the scanning electron microscopy (SEM) microphotographs of the polished N80 steel sample (magnification ×1,000). Figure 14b shows the SEM microphotographs of N80 steel (magnification ×1,000) when exposed to 15% HCl solution at room temperature in the absence of inhibitors. Figure 14b also shows that the steel surface appears to be very rough in the absence of inhibitors. This is due to the formation of uniform flake-type corrosion products on the metal surface. No pitting and other separate phases are visible in microphotograph. Figure 14c,d shows the microphotographs of the metal surface when exposed to the acid medium in the presence of 200 ppm of [PAMTOI] and [MMTOI], respectively, at the same magnification. Comparing the microphotograph of the steel surface without an inhibitor with the photographs of the exposed surface in the presence of inhibitors, the steel was found to be covered with a semiglobular-type protective film of compounds uniformly spread over the surface. The protective film is formed due to adsorption of the inhibitor molecules on the N80 steel surface.

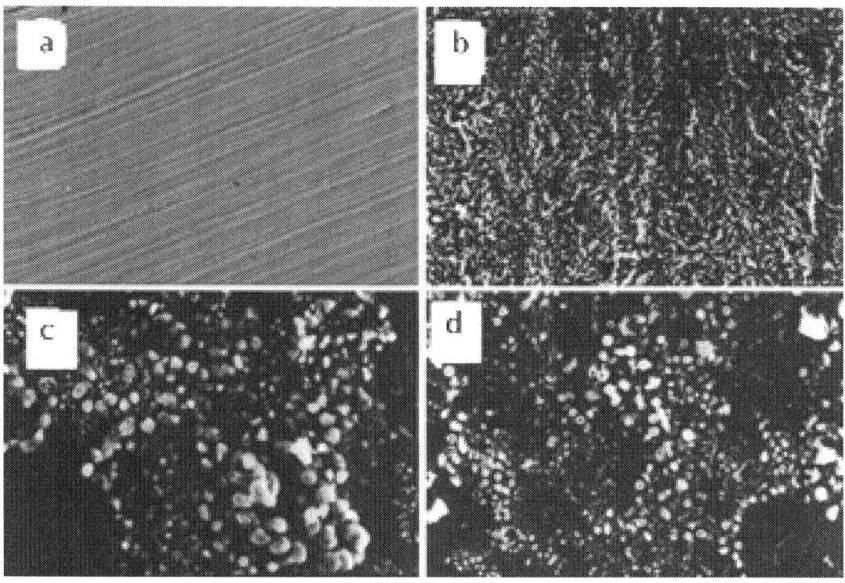

Figure 14: SEM images. (a) Polished sample. (b) Sample with 15% HCl. (c) Sample with 200 ppm of [PAMTOI]. (d) Sample with 200 ppm of [MMTOI].

EXPERIMENTAL

N80 Steel Sample Preparation

The corrosion studies were performed on mild steel samples with the following composition (wt.%): C, 0.31; Mn, 0.92; Si, 0.19; P, 0.01; S, 0.008; Cr, 0.20; and Fe, remainder. The N80 steel coupons having the size dimension 3.0 cm × 3.0 cm × 0.1 cm were mechanically cut and abraded with emery papers of different grades (120, 220, 400, 600, 800, 1,500, and 2,000) for weight loss experiment. For electrochemical measurements, mild steel coupons having the dimension 1.0 cm × 1.0 cm × 0.1 cm were mechanically cut and abraded in the same manner as before, with an exposed area of 1 cm²(the rest covered with araldite resin) with a 3-cm-long stem. Prior to the experiment, specimens were washed with distilled water, degreased in acetone, dried, and stored in a vacuum desiccator.

Test Solutions

For weight loss study, the test solutions (15% HCl, wt.%) were prepared by dilution of analytical-grade 37% HCl (Rankem, Faridabad, India), and the required concentrations of inhibitors were calculated before the 15% HCl solutions have been made up. The concentrations of the studied inhibitors ranging from 20 to 200 ppm by weight in 15% HCl were prepared. All solutions were prepared in double-distilled water.

CONCLUSIONS

[PAMTOI] and [MMTOI] both act as good corrosion inhibitors for the corrosion of N80 steel in 15% HCl solution. The inhibition efficiency values increase with the inhibitor concentration, but decreases with increasing temperature for the corrosion of N80 steel in 15% HCl solution. The adsorption of [PAMTOI] and [MMTOI] on the N80 steel surface obeys the Langmuir adsorption isotherm. The variation in the values of a and c (Tafel slopes) and the minor negative shift in the values of the corrosion potential (E_{corr}) indicate that both tested inhibitors are of mixed type but predominantly control the cathodic reactions. EIS measurements show that R_{ct} increases and C_{dl} decreases in the presence of inhibitors. The FTIR data for the synthesized product and those for the metal surface product suggested the adsorption of the inhibitor molecules on the surface of N80 steel.

AUTHORS' CONTRIBUTIONS

MY carried out the data treatment and drafted the manuscript. US carried out the synthesis of the inhibitors and the gravimetric and electrochemical measurements. PY studied the effect of temperature and carried out the thermodynamic parameter calculations. All authors read and approved the final manuscript.

ACKNOWLEDGEMENTS

Financial assistance from the Indian School of Mines, Dhanbad, under the 'Faculty Research Scheme' to M. Yadav is gratefully acknowledged.

REFERENCES

1.   Vishwanatham S, Haldar N (2008) Corros Sci. 50:2999
2.   Abd El-Maksoud SA, Fouda AS (2005) Mater Chem Phys. 93:84
3.   Migahed MA, Nassar IF (2008) Electrochim Acta. 53:2877
4.   Mohammed AA, Ibrahim M (2011) Corros Sci. 53:873
5.   Nataraja SE, Venkatesha TV, Manjunatha K (2011) Corros Sci. 53:2651
6.   Ghareba S, Sasha O (2010) Corros Sci. 52:2104
7.   Aljourani J, Raeissi K, Golozar MA (2009) Corros Sci. 51:1836
8.   Emregal CK, Mustfa H (2006) Corros Sci. 48:797
9.   Muralidharan S, Quraishi MA, Iyer SVK (1995) Corros Sci. 37:1739
10. Bentiss F, Lebrini M, Lagrenee M, Traisnel M, Elfarouk A, Vezin H (2007) Electrochim Acta. 52:6865
11. Cruz J, Martinez R, Genesca J, Garcia-Ochoa E (2004) J Electroanal Chem. 566:111
12. Khaled KF, Babic-Samradzija K, Hackerman N (2005) Electrochim Acta. 50:2515
13. Roberge PR (1999) Corrosion inhibitors: handbook of corrosion engineering New York: McGraw-Hill
14. Olivares-Xometl O, Likhanova NV, Domínguez-Aguilar MA, Arce E, Dorantes H, Arellanes-Lozada P (2008) Mater Chem Phys. 110:344
15. Quartarone G, Battilana M, Bonaldo L, Tortato T (2008) Corros Sci. 50:3467
16. Ebenso EE (2003) Bull Electrochem 19:209
17. Gardner GS, Saukaitis AJ (1957) US Patent 2:807,585
18. Quraishi MA, Ahamad I, Singh AK, Shukla SK, Lal B, Singh V (2008) Mater Chem Phys. 112:1035

19. Abdulghani AJ, Abbas NM (2011) Bioinorg Chem Appl.

20. Putilova IN, Balezin SA, Barannik VP (1960) Metallic corrosion inhibitors New York: Pergamon

21. Emregül KC, Hayvalı M (2006) Corros Sci. 48:797

22. Shorky H, Yuasa M, Issa SI, El-Baradie RMHY, Gomma GK (1998) Corros Sci. 40:2173

23. Mu GN, Zhao TP, Liu M, Gu T (1996) Corrosion. 52:853

24. Schweinsberg DP, George GA, Nanayakkara AK, Steiner DA (1988) Corros Sci. 28:33

25. Olivares O, Likhanova N, Gomez B (2006) Appl Surf Sci. 252:2894

26. Al-Juaid SS (2011) Chemistry and Technology of Fuels and Oils. 47:58

27. Dehri I, Ozcan M (2006) Mater Chem Phys. 98:316

28. Behpour M, Ghoreishi SM, Soltani N, Salavati-Niasari M, Hamadanian M, Gandomi A (2008) Corros Sci. 50:2172

29. Lebrini M, Traisnel M, Lagrene M, Mernari B, Bentiss F (2007) Corros Sci 50:473

30. Ozcan M (2008) J Solid State Electrochem. 12:1653

31. Ashassi-Sorkhabi H, Majidi MR, Seyyedi K (2004) Appl Surf Sci. 225:176

32. Li XH, Deng SD, Fu H (2009) Corros Sci. 51:1344

33. Lebrini M, Lagrenée M, Vezin H, Traisnel M, Bentiss F (2007) Corros Sci. 49:2254

34. Rosenfield IL (1981) Corrosion inhibitors New York: McGraw-Hill

35. MaCafferty M, Hackerman N (1972) J Electrochem Soc. 119:146

36. Silverstein RM, Bassler GC, Morrill TC (1991) Spectrometric identification of organic compounds New York: Wiley

Floating Layer Formation, Foaming, and Microbial Community Structure Change in Full-scale Biogas Plant Due to Disruption of Mixing and Substrate Overloading

Tobias Lienen[1], Anne Kleyböcker[1], Manuel Brehmer[2],
Matthias Kraume[2], Lucie Moeller[3], Kati Görsch[3], and
Hilke Würdemann[1]

[1]GFZ German Research Centre, Telegrafenberg, 14473 Potsdam, Microbial GeoEngineering, Germany

[2]TechnischeUniversität Berlin, Fachgebiet Verfahrenstechnik, Ackerstraße 71-76, Berlin 13355, Germany

[3]Environmental and Biotechnology Centre, UFZ Helmholtz Centre for Environmental Research, Permoserstrasse 15, Leipzig 04318, Germany

ABSTRACT

Background

The use of biogas as renewable resource of energy is of growing interest. To increase the efficiency and sustainability of anaerobic biogas reactors, process failures such as overacidification, foaming, and floating layers need to be investigated to develop sufficient countermeasures and early warning systems to prevent failure.

Methods

Chemical, rheological, and molecular biological analyses were conducted to investigate a stirring disruption in a full-scale biogas plant.

Results

After the agitation system was disturbed, foaming and floating layer formation appeared in a full-scale biogas plant fed with liquid manure and biogenic waste. Rheological characterizations and computational fluid dynamics (CFD) revealed a breakdown of the circulation within the reactor and a large stagnation zone in the upper reactor volume. Molecular biological analyses of the microbial community composition in the floating layer showed no differences to the digestate. However, the microbial community in the digestates changed significantly due to the stirring disturbances. Foam formation turned out to be a consequence of overloading due to excessive substrate supply and disturbed mixing. The subsequent increase in concentration of both acetic and propionic acids was accompanied by foaming.

Conclusions

Effective mixing in full-scale biogas plants is crucial to avoid foaming and floating layers and to enhance sustainability. Disturbed mixing leads to process imbalances and significant changes in the microbial community structure. Additionally, controlled feeding might help prevent foam formation due to overloading.

BACKGROUND

Anaerobic cofermentation of biogenic wastes to produce biogas is of growing interest to generate renewable energy and to reduce greenhouse gas emissions. The production of biogas by biogenic wastes is a decentralized technology and contributes toward the renewable energy turnaround in Germany. Biogas plants are often affected by process failures such as overacidification or foam and floating layers that reduce the efficiency of plants. A breakdown of the biogas production process because of process failures leads to enormous economical loss and deteriorated sustainability [1]. Selection of suitable substrates and proper mixing are important challenges in the biogas production industry to avoid failures of the process. Feeding with profitable but unsuitable substrates may lead to serious process interruptions. An overacidification event is often caused by substrate overloading and accumulation of volatile fatty acids (VFA) [2]. Kleyböcker et al. [3] developed two early warning indicators in terms of overacidification. The first indicator (EWI-VFA/Ca) is characterized by the relation of VFA to Ca^{2+} and was shown to provide a warning 5 to 7 days before an overacidification appeared. The warning is indicated by a two- to threefold increase of values. The second early warning indicator (EWI-PO_4/Ca) is characterized by the relation of PO_4^{3-} to Ca^{2+}. Moeller et al. [4] showed a correlation between the fed substrates as well as inadequate plant management and the formation of foam in biogas plants. Foaming can be caused by high concentrations of VFA, surface-active compounds, detergents, proteins, and high nitrogen concentrations as well as organic overloading [4]. In contrast, the formation of floating layers is mainly triggered by inadequate mixing and feeding of fibrous substrates [5]. Besides the chemical and physical characteristics of the substrates and the plant management, foam and floating layers are also promoted by growth of filamentous bacteria [6, 7]. Most of the studies regarding the filamentous bacteria were done in activated sludge treatment plants and anaerobic digesters fed by sewage sludge. In these systems, mainly *Gordonia* spp. and *Microthrix parvicella* were identified as foam causers [8, 9]. Using their filamentous structure, the microorganisms trap biogas bubbles, which transfer them to the surface. The hydrophobic cell surface promotes and stabilizes the foam as well as the release of hydrophobic substances. Foaming and the formation of floating layers may cause serious damage in biogas plants [10]. The active volume of the digester is reduced leading to an inefficient gas recovery. Furthermore, gas mixing devices may be blocked, gas pipelines may be fouled, and even the roof may be damaged by the pressure of the foam or floating layer. In addition, the economical costs of energy loss, manpower overtime, and cleaning costs have to be taken into consideration [11]. Although several studies on foam formation in anaerobic digesters treating activated sludge were published [8,12], formation of floating layers and foam in anaerobic digesters fed with biogenic waste and the impact of mixing procedures in full-scale biogas plants as well are still rarely investigated. Most importantly, information about the chemical composition and microbial community structure of the floating layer is lacking due to the difficulties of sample collection in full-scale digesters. The objective of the study presented in this paper was to investigate the alterations in rheological, chemical, and microbiological parameters during an agitator disruption accompanying floating layer and foam formation in a full-scale biogas plant treating biogenic waste. In addition, the repair of a broken stirring paddle allowed for analyzing a floating layer chemically and microbiologically. Furthermore, two early warning indicators for overacidifications were used to investigate the process performance in the floating layer. The results of this study give further insight into the improvement of the efficiency of the biogas production process as well as understanding of the complex microbial community composition.

METHODS

Biogas Plant Scheme, Process Operation, and Sampling

The full-scale biogas plant was run as a two-stage plant consisting of two hydrolysis reactors H1 and H2 (520 m³) operated at 30°C and two methanogenic reactors R1 and R2 (2,300 m³) operated at 37°C

(Figure 1). The hydraulic retention time was kept within the range of 26 days. Liquid manure (50,000 t/a) and a highly variable combination of biogenic wastes (30,000 t/a) from the fish industry and oil from fat separators as well as creamery and slaughterhouse waste were fed as substrates. R1 and R2 were charged in turn every 4 h by 15 m³ substrate. Mixing was conducted by continuous stirring with paddles in two different heights. The methanogenic reactors were operated at an organic loading rate between 2 and 2.5 kg VS m^{-3} day^{-1}. The produced biogas consisted of 60% to 65% CH_4. The biogas production was about 4.4 million m³/a while the digestate residues yielded 61,000 t/a. The methanogenic reactors R1 and R2 were monitored from the start of an agitator breakdown in these reactors over a period of 6 months. In month 5, stirring was reconstituted. While the agitator motor in R1 broke down completely, R2 was affected by one broken stirring paddle (Figure 2a). During the first month, foam was observed in both reactors. The foam disappeared after 1 day without intervention. Additionally, a sample of a floating layer was directly collected during the reparation process of the broken paddle in R2 in month 5 (Figure 2b). In the first month, digester samples were withdrawn at the drain at the bottom of the two reactors bi-weekly. Afterward, sampling was conducted monthly.

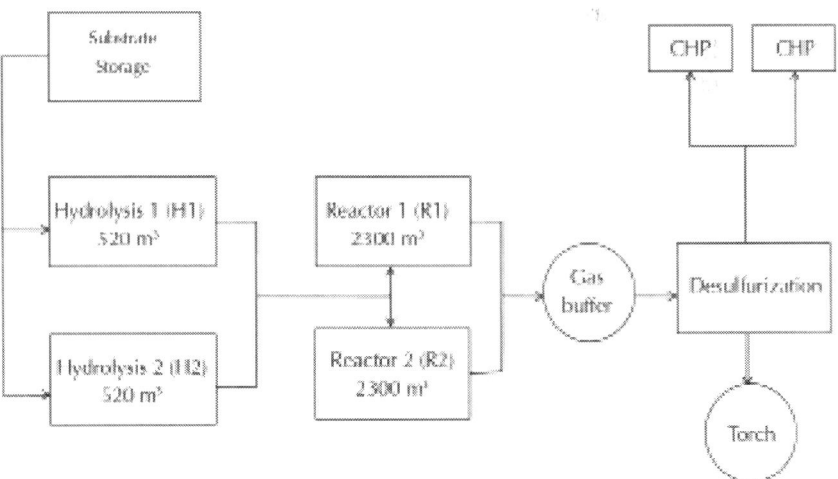

Figure 1: Scheme of the full-scale biogas plant. CHP, combined heat, and power plant.

Figure 2: Broken stirring paddle and floating layer in R2. (a) Exchange of broken stirring paddle in R2. (b) Floating layer in R2.

Chemical Analyses

For the total solids (TS) and the volatile solids (VS), about 50 g of the samples was dried at 105°C in a Memmert drying chamber (Memmert GmbH, Hannover, Germany) for 24 h and then burned at 550°C (Nabertherm Controller B170; Nabertherm GmbH, Lilienthal/ Bremen, Germany). The weight of the samples was determined using a Sartorius CP220S-OCE balance (Sartorius AG, Goettingen, Germany) (scale ± 0.01 g). The TS and VS were analyzed according to German guideline DIN 38409–1 [13]. For the chemical parameters and for DNA extraction, the samples were centrifuged 10 min at 12,857×g to pelletize solid substances. Pellets were transferred into 1.5 mL tubes and stored at -20°C to extract DNA afterward. The supernatant was centrifuged again for 10 min at 12,857×g. The volatile fatty acids (VFA), calcium (Ca^{2+}), and phosphate (PO_4^{3-}) concentrations were measured photometrically (Hach-Lange DR2800, Hach-Lange cuvette tests LCK 365, 327, and 350; Hach Lange GmbH, Düsseldorf, Germany) in the second supernatant. Additionally, the two early warning indicators EWI-VFA/Ca and EWI-PO_4/Ca for overacidifications and process imbalances [3] were tested. The nitrogen and organic/inorganic carbon concentrations were detected in homogenized (using a common hand-held blender) samples by a TOC-VCSH/CSN analyzer containing a

TN-unit (Shimadzu, Nakagyo-ku, Kyoto, Japan). The crude protein concentration was determined according to the method of Dumas [14] with a modified determination method of total nitrogen concentration (*videsupra*). The factor volatile organic acids/total inorganic carbonate buffer (VOA/TIC) determines the buffer capacity of the biogas producing system [15]. The VOA/TIC analysis was carried out according to the Nordmann method [16] using 20 mL of centrifugation supernatant (20 min, 5,300 rpm, and 20°C; Beckman Avanti 30 centrifuge, Brea, CA, USA). The 10-mL samples were filtered through a membrane filter (nylon, 0.45 µm, Pall Corp, Port Washington, NY, USA) for further analysis. The ammonia nitrogen concentration was measured using the spectrophotometric test kit Spektroquant® (Merck KGaA, Darmstadt, German) with photometer MuliLab P5 (WTW, Weilheim, Germany). The concentrations of volatile organic acids (acetic acid, propionic acid, and butyric acid) were measured by use of HPLC (Shimadzu: VA 300/7.8 nucleogelion 300 OA column, 0.01 N H_2SO_4 as eluent, and the detector RID-10A). Water-elutable elements were determined by ICP-AES (according to EN ISO 11885, Spectro, Kleve, Germany).

Rheological Characterization

Due to the complex structure of the substrate, the viscosity was determined by using different measuring systems. Three relative methods - a pipe viscosimeter, a paddle mixer, and a ball measuring system - were compared with two standardized methods: the double gap and the cylinder measuring system. In the double gap measuring system, an additional static cylinder is in the center and the rotating measuring cylinder is hollow. Thus, two spaces are created between the inner stator and rotor as well as between the rotor and the outer wall. The calculations of the apparent viscosity η_s and the shear rate $\dot{\gamma}$ are summarized in Table 1. The measurement systems used to describe the rheology differ in their measuring range. Apart from the results of the double gap measuring system, the viscosity showed the expected dependence on the shear rate. The higher values of this measuring system were due to the large friction of the particles contained in the substrate at the stator because of the small gap width. Based on this comparative study, the cylinder measuring system was chosen for further viscosity monitoring because it was easy to handle and required a comparably low sample volume. The rheological characterization of the substrates was performed over a period of more than 1 year, including the period when the agitator system was disrupted. The shear thinning characteristics of the substrate are described by the power law equation $\eta_s = K.\dot{\gamma}^{n-1}$, wherein K stands for the Ostwald factor and η for the shear rate exponent. Based on this equation, the flow regime of the biogas plant with its typical rheological characteristic was determined using the computational fluid dynamics (CFD) software CCM+. The cylindrical model, with a diameter of 13 m and a height of 15.3 m, had a two-stage central mixer in line with the plant. The height was equal to the liquid level; therefore, the ceiling of the cylinder and the boundary layer between the liquid and the gas phase was modeled assuming slip conditions. The lower agitator had a diameter of 4.2 m and was installed 4 m above the ground. A second agitator with a diameter of 2 m was placed at a height of 12 m above the ground. Both impellers were attached to the same shaft. The resulting liquid volume was displayed in the CFD software with a polyhedron grid and approximately 4 million cells. This calculation area was broken down into a stagnant self-contained cylinder and a rotating cylinder. To model the transfer of mass, momentum, energy, and other physical quantities between these two regions, the 'indirect' interface was used. For consideration of the agitator moving, the so-called moving reference frame model was applied. The rotation frequency was 42 rpm and steady state conditions were assumed. To compare both relevant process conditions, with and without a disrupted agitator, a second simulation with disrupted stirring was performed, while the boundary conditions were kept constant and the lower agitator blades were removed.

Table 1: The calculation of the apparent viscosity η_s and the shear rate $\dot{\gamma}$ [17], [18]

Measuring system	Equation	Constant

Double gab measuring system	$$\tau = \frac{1+\delta^2}{(\delta^2 \cdot R_3^2 + R_2^2)} \cdot \frac{M}{4000 \cdot \pi L C_L}$$	CL = 1.10
		= 1.0245
		RMKi = 20.25 mm
	$$\dot{\gamma} = \omega \cdot \frac{1+\delta^2}{\delta^2 - 1}$$	RMKa = 21.00 mm
		L = 78.7 mm
Cylinder measuring system	$$\tau = \frac{1+\delta^2}{2000 \cdot \delta^2} \cdot \frac{M}{2\pi L \cdot R_i^2 \cdot C_L}$$	CL = 1.10
		= 1.0848
	$$\dot{\gamma} = \omega \cdot \frac{1+\delta^2}{\delta^2 - 1}$$	RMKi = 13.33
		L = 40.003 mm
Ball measuring cell	$$\tau = C_{SS} \cdot M$$	$C_{SS} = 15.0 \frac{Pa}{mNm}$
	$$\dot{\gamma} = C_{SR} \cdot n$$	$C_{SR} = 0.427 \frac{1/min}{s}$ $d_K = 12.0$ mm
		dK = 12.0 mm
Paddle mixer [17]	$$Ne = \frac{P}{p \cdot n^3 \cdot d^5} = \frac{C_{lam}}{Re} C_{turb}$$	CMO = 10.94
		Clam = 189.6
	$$\dot{\gamma} = C_{MO} \cdot n$$	Cturb = 9.4
		dR = 30 mm
		dB = 143 mm
Pipe viscosimeter [18]	$$\tau_W = \frac{d \cdot \Delta p}{4 \cdot L}$$	Lpipe = 2,500 mm
	$$\left(\frac{dw}{dr}\right)_W = \frac{3n'+1}{4n'} \cdot \frac{8 \cdot w_m}{d}$$	dpipe = 43.2
	$$n' = \frac{d\ln\left(\frac{d \cdot \Delta p}{4 \cdot L}\right)}{d\ln \frac{(8 \cdot w_m)}{d}}$$	

Lienen *et al.*

Lienen *et al. Energy, Sustainability and Society* 2013 **3**:20, doi:10.1186/2192-0567-3-20

DNA Extraction and PCR-DGGE Analysis

To compare the diversity in the microbial community compositions, the total genomic DNA was extracted from 350 mg of the pellets using the MP Fast DNA Spin Kit for Soil according to the manufacturer's instructions. The partial 16S rRNA genes (566 bp) of the bacterial community were amplified by polymerase chain reaction (PCR) in 50 µL reactions with 1 µL of 1:10 diluted template using the primer pair 341 F-GC/907R [19,20] (94°C 2:45 min, 94°C 0:45 min, 56°C 0:45 min, 72°C 0:50 min, 72°C 30 min, 40 cycles). Amount of 50 µL of reactions was mixed containing 5 µL 10× reaction buffer (Genecraft, Lüdinghausen, Germany), 6 µL dNTPs (10 mM, Fermentas, Thermo Fisher Scientific, Waltham, MA, USA), 3 µL $MgCl_2$ (50 mM, Genecraft), 3 µL forward primer (10 mM), 3 µL reverse primer (10 mM), 0.4 µL BSA (20 mg/mL, Fermentas), 0.3 µL Taq polymerase (5 u/µL, Genecraft), 28.3 µL RNA/DNA-free water (Fermentas), and 1 µL of 1:10 diluted template. Amplicons were purified subsequently using the Fermentas GeneJET PCR Purification Kit (Fermentas, Thermo Fisher Scientific, Waltham, MA, USA) and the amplicon concentration was determined fluorimetrically (BMG Labtech FLUOstar OPTIMA; BMG LABTECH GmbH, Allmendgruen, Ortenberg, Germany) by labeling the DNA with Quant-iTPicoGreen (Invitrogen, Darmstadt, Germany). Denaturing gradient gel electrophoresis (DGGE) was performed afterward with equal concentrations of amplicons and a gradient of 35% to 65% urea and 6% acrylamide (BioradDCode System, Munich, Germany). The DGGE gel ran for 17 h at 110 V and 60°C. Bands of interest were excised and transferred into a 0.5-mL tube. Amount of 50 µL of sterile H_2O was added and removed directly to wash the gel pieces. Afterward, 30 µL sterile H_2O was added. The tube was shaken for 1 h at 37°C to recover the DNA out of the gel. Reamplification was carried out using 4 µL template of recovered DNA and the primer pair 341 F/907R (94°C 1:30 min, 94°C 0:30 min, 56°C 0:30 min, 72°C 0:30 min, 72°C 10 min, 30 cycles). PCR products were purified using the Avegene gel/PCR DNA fragments extraction kit (MSP KOFEL, Zollikofen, Switzerland), and the DNA concentrations were measured fluorimetrically according to the procedure mentioned above. The PCR products were sent in and sequenced by GATC Biotech AG (Jakob-Stadler-Platz 7, Konstanz, Germany). Sequences were edited using the BioEdit Sequence Alignment Editor version 7.0.5.3 [21]. Basic Local

Alignment Search Tool (BLAST) [22] was used for sequence similarity check, and the taxonomic assignment was done by Ribosomal Database Project (RDP) using the RDP Classifier [23]. Based on the DGGE profiles, a graphical representation of the bacterial community evenness was set by using Pareto-Lorenz (PL) distribution curves [24] as previously described by Wittebolle et al. [25]. GelQuant.NET software provided by biochemlabsolutions.com was used to determine the band intensities. The band intensities for every DGGE lane were ranked from high to low and the cumulative band intensities were used as the y-axis. The cumulative normalized number of bands was set as the x-axis. Evaluation of the curves was conducted by comparison to a vertical 20% x-axis line. The theoretical perfect evenness line was set as 45° diagonal.

Microscopy

The floating layer sample was analyzed by bright field microscopy at a ×100 magnification (Zeiss Axio Imager M2; Carl Zeiss, Oberaue 3, Jena, Germany). Therefore, the sample was added to a drop of water on an object slide and viewed microscopically afterward.

RESULTS AND DISCUSSION

Foam Formation Related to Substrate Overloading

Formation of foam was observed in both methanogenic reactors on day 12 in the first month after high organic loading and agitator break-down. Since the monitoring started when the stirring was disturbed, chemical analyses for the period before were not accessible. Besides breakdown and disturbance of stirring as a cause for the foam formation during the first month in both methanogenic reactors, changes in the substrate mix also have to be considered (Figure 3). The hydraulic retention time of the hydrolytic reactors was 11.2 days, and considering that the substrate feeding occurred in the period of 12 days before foaming in the methanogenic reactors, a peak in the total quantity of fed substrates was applied by the operator. After breakdown of agitation foaming occurred. The chemical characterization of the digestate from the time period before and after foaming is shown in Table 2. The increase of VOA/TIC values amounted to 0.08 and 0.06 during 4 days in both methanogenic reactors, respectively. Accordingly, the concentrations of VOA rose from 1,600 mg L^{-1} (R1) and 1,650 mg L^{-1} (R2) to 2,700 mg L^{-1} (R1) and 2,500 mg L^{-1} (R2), respectively. The concentration of acetate increased in R1 from less than 1 mg L^{-1} to 400 mg L^{-1} and in R2 from 100 mg L^{-1} to 200 mg L^{-1}. Moreover, in only one sample, 66 mg L^{-1} of propionic acid was detected in R2 1 day after the foaming occurred. The acetic acid concentration decreased in R1 5 days after foaming, while it was 2.5-fold increased in R2. Presumably, high feeding together with disturbed mixing led to an overloading of the reactors. Some of the fed substrates such as fish and slaughterhouse waste as well as easily degradable substrates such as sugar beet molasses are well known to favor foaming [26,27]. Although the VOA/TIC is specific for each biogas plant, a sudden change in the VOA/TIC curve indicates a process disturbance. The accumulation of intermediates of the biogas producing process is known to be a consequence of failure of the microbial process that can be due to organic overloading [28]. One and two days before foaming appeared, the early warning indicators EWI-VFA/Ca and EWI-PO$_4$/Ca increased by a factor of 2 (Figure 4). According to Kleyböcker et al. [3], the increase indicates a warning in terms of overacidification and overloading. Because the overloading is regarded as a reason for foaming, the EWI-VFA/Ca and EWI-PO$_4$/Ca warned in terms of foam formation as well. The ammonia nitrogen concentration was more or less constant. One day after foaming, the crude protein concentration increased considerably in R1, while it stayed almost stable in R2. Furthermore, the protein concentration decreased considerably in both methanogenic reactors 5 days after foaming. Concerning water eluable elements, diverse trends were observed. Calcium and magnesium concentrations were higher in both fermenters 1 day after foaming than in the period before foaming. The calcium concentration decreased again in both methanogenic reactors 5 days after foaming. The magnesium concentration had an opposite tendency in both fermenters; it rose further in R1 while sinking in R2. Nickel was detected with a concentration of about 20 mg L^{-1} in both methanogenic reactors 1 day after foaming. All other element concentrations were either stable or only slightly higher in the period after foaming (Table 2). The role of water eluable elements in the foam formation in biogas plants has not been studied so far. Nevertheless, the experience from fermentation processes in digestive systems of ruminants may help to understand biogas systems. Miltimore [29] found that calcium, nickel, and zinc were associated with the bloat of ruminants, whereas magnesium had no relation to foaming in the rumen. The increased calcium and nickel concentrations during the foam formation most likely resulted from the variances in the substrate mix. Moreover, there was a considerable drop in the biogas production rate after the foaming (Figure 3), also indicating a process imbalance. Unfortunately, no sampling of the foam from the reactor surface was possible to analyze the chemical composition of the foam in order to confirm these assumptions.

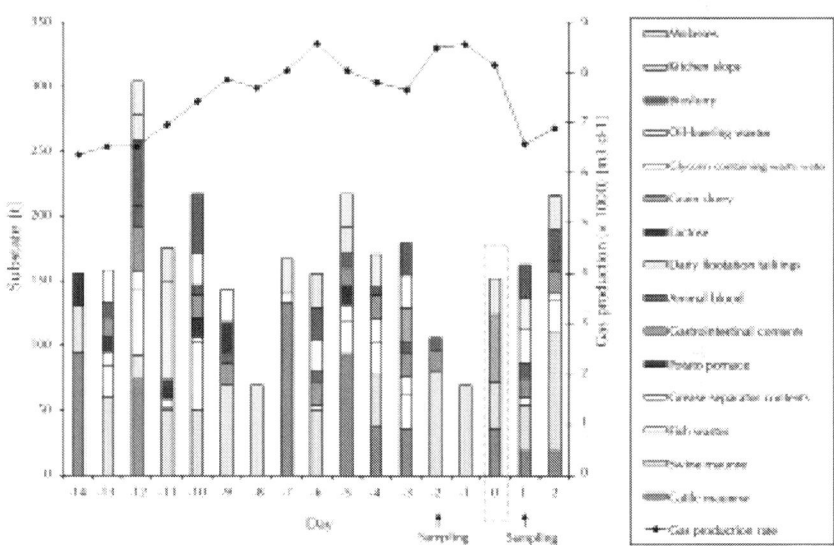

Figure 3: Substrate feeding ratios during and before the foaming period in month 1. The dashed lined rectangle indicates the day when both methanogenic reactors foamed (day '0'). High substrate loading was detected 12 days before the foam appeared.

Table 2: Chemical characterization of the digestate of the methanogenic reactors R1 and R2 during foaming

	Two days before foaming		One day after foaming		Five days after foaming	
	R1	R2	R1	R2	R1	R2
VOA/TIC (-)	0.093 ± 0.006	0.098 ± 0.008	0.170 ± 0.023	0.155 ± 0.002	0.134 ± 0.036	0.0173 ±0.038
Total carbon (g L-1)	16.42 ± 0.86	15.55 ± 0.69	18.63 ± 1.23	13.67 ± 0.30	15.87 ± 0.58	15.80 ± 0.66
Total organic carbon (g L-1)	12.23 ± 0.76	11.44 ± 0.64	14.04 ± 0.52	9.793 ± 250	11.87 ± 0.51	11.75 ± 0.66
Total nitrogen (g L-1)	4.63 ± 0.24	4.57 ± 0.19	5.14 ± 0.14	4.60 ± 0.15	4.21 ± 0.05	4.28 ± 0.05
Ammonia nitrogen (g L-1)	2.61 ± 0.03	2.69 ± 0.02	2.83 ± 0.01	2.85 ± 0.01	2.94	2.98
Crude protein (g L-1)	12.67	11.73	14.42	10.97	7.95	8.16
Volatile organic acids:						
Acetic acid (mg L-1)	<1	126	407	207	215	551
Propionic acid (mg L-1)	<1	<1	<1	66	<1	<1
Butyric acid (mg L-1)	<1	<1	<1	<1	<1	<1
Water-eluable elements						
Calcium (mg L-1)	43.4	56.5	54.3	86.0	36.3	61.8

Iron (mg L-1)	5.33	5.56	5.20	4.88	4.26	6.91
Magnesium (mg L-1)	4.57	5.81	6.30	12.5	11.4	9.63
Nickel (mg L-1)	<4	<4	21.4	20.8	<4	<4
Phosphorus (mg L-1)	132	145	161	139	151	151
Potassium (mg L-1)	1,703	1,687	1,682	1,659	1,755	1,690
Sulfur (mg L-1)	46.6	43.2	56.9	43.1	47.3	49.6

In case of multiple determinations, standard deviations are presented.

Lienen et al.

Lienen et al. Energy, Sustainability and Society 2013 3:20, doi:10.1186/2192-0567-3-20

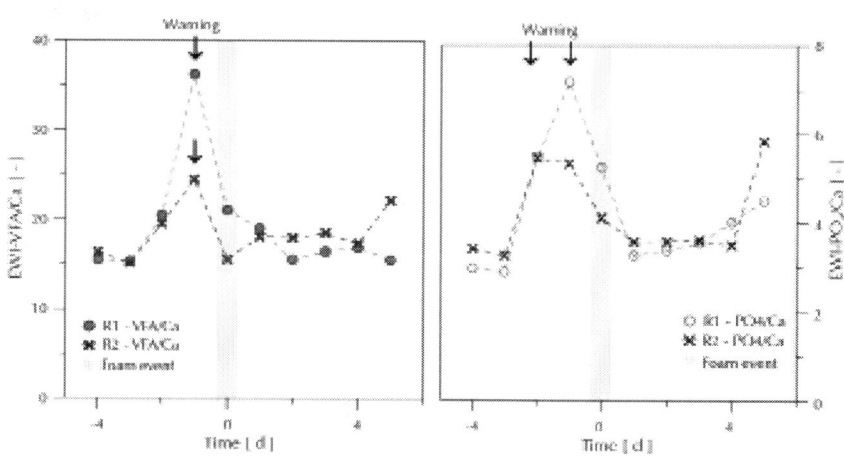

Figure 4: Early warning indicators EWI-VFA/Ca and EWI-PO$_4$/Ca. Before, during, and after the foam event in the reactors R1 and R2 in month 1. Both indicators increase 1 to 2 days before foaming appeared. The increase is interpreted as a warning in terms of overloading. The overloading is regarded as reason for the formation of foam.

Stagnant Zones, Sedimentations, and Process Imbalances Due to Stirring Disturbances

The analysis of the rheology showed that the values for the Ostwald factor and the shear rate fluctuated considerably within the monitoring period of 6 months (Figure 5). Surprisingly, no significant changes of the rheological characteristics were detected during the period without stirring. Based on this rheological result, an Ostwald factor of 0.125 $Pas^{0.53}$ and a flow exponent of 0.53 were chosen for the CFD simulations. However, the power law approach was used for a shear rate range from 0.01 to 700 s^{-1} only. Outside this range, a constant viscosity of either 0.01 Pas or rather 4.03 was assumed. The open jet of the substrate supply was not considered in the simulation. The thoroughly mixed volume generated by the open jet was estimated as 5% to 10% of the whole reactor. With a feeding interval of 4 h, the mixing process by substrate dosage was negligible. Under undisturbed mixing conditions, the numerical simulations showed a good mixing for the studied biogas plant (Figure 6a). However, the partial stirring did not prevent the formation of floating layers in modeling the flow velocity since especially in the upper reactor volume, stagnant zones appeared (Figure 6b). The stagnation zone with a velocity less than 0.05 $m \cdot s^{-1}$ was limited to a volume of 52.6 m^3 and represented 2.4% of the whole reactor. After breakdown of the lower agitator, large stagnation zones were formed, especially in the upper reactor volume with a volume of 487 m^3 that was about 21.8% of the whole reactor and the whole flow field collapsed (Figure 6b). The concentrations of TS and VS in the monitoring period covering the time from agitator breakdown to the restart of stirring after 5 months differed slightly between R1 and R2 with differences from 3 to 4 g L^{-1} (Figure 7). The complete agitator downtime in R1 most likely led to sedimentations. The small volume of the introduced substrate was not sufficient to blend the reactor, and sedimentation occurred resulting in a 10% higher TS value and 13% higher VS value in R1 than in R2 in which no or less sedimentation occurred due to the partial mixing (Figure 7). The substrates were concentrated at the bottom of R1 and directly withdrawn at the drain during sampling which led to an overestimation of the average TS and VS values. In contrast to the complete stirring downtime in R1, the substrates in the partially stirred R2 were better distributed leading to a more sufficient degradation of organic matter and lower TS and VS values. The average pH values of the digestates of the methanogenic reactors were slight alkaline with a value of about 8 (Figure 7). The gas production rate was decreased by 16% to 36% during the mixing disturbances (Figure 8). Unfortunately, the biogas production of the two reactors was quantified together so that the comparison of the biogas production rate was not feasible. Analysis of the VFA concentrations in the methanogenic reactors showed increasing values after the agitators were broken (Figure 8). In the first 2 months, the VFA in both reactors nearly doubled from about 2,500 mg L^{-1} to 3,900 mg L^{-1} and the EWI-VFA/Ca increased two to three times. According to Kleyböcker et al. [3] the two- to threefold increase in the EWI-VFA/Ca indicates a process imbalance. However, the VFA concentrations in the hydrolytic reactors increased as well. In the first month, the VFA concentrations in both hydrolytic reactors reached a value of about 18,500 mg L^{-1} and increased in the second month up to more than 23,000 mg L^{-1}. Afterward, the concentrations decreased constantly to less than 14,000 mg L^{-1} in month 5. Correspondingly, from month 3 onward, the VFA concentrations in the two methanogenic reactors decreased and stabilized at a value of around 1,200 mg L^{-1}, and the EWI-VFA/Ca also indicated process stabilization. The small intensity of stirring very likely favored the stabilization process. Stroot et al. [30], Gomez et al. [31], and Kaparaju et al. [32] also observed process stabilization due to gentle and/or minimal mixing (intermittent mixing) after high organic loading. It is probable that the increased concentrations of VFA in the methanogenic reactors were mainly caused by higher VFA concentrations in the fed substrate combined with an insufficient distribution of the substrate due to the stirring disturbances. Rojas et al. [33] revealed a decreasing performance of the biogas process in a reactor without stirring and related it to the insufficient contact between substrate and microorganisms. After restart of the agitator in month 5, the VFA concentrations in the two methanogenic reactors remained on a low level although the VFA concentrations in the hydrolytic stage increased again to more than 24,000 mg L^{-1}. In month 5, a floating layer sample was collected from the surface of R2 during the exchange of the broken stirring paddle. The TS and VS values of the floating layer were increased four- to fivefold related to the digestate, and the VFA concentration was twofold higher than in the digestate at a value of 2,200 mg L^{-1} (data not shown). The EWI-VFA/Ca was increased fivefold compared to the digestate, whereas the EWI-PO_4/

Ca was increased threefold and indicated a process imbalance[3]. The accumulation of organic components in the layer and very limited exchange of intermediates within the layer led to an accumulation of VFA and therefore the microbial degradation process was inhibited in the floating layer.

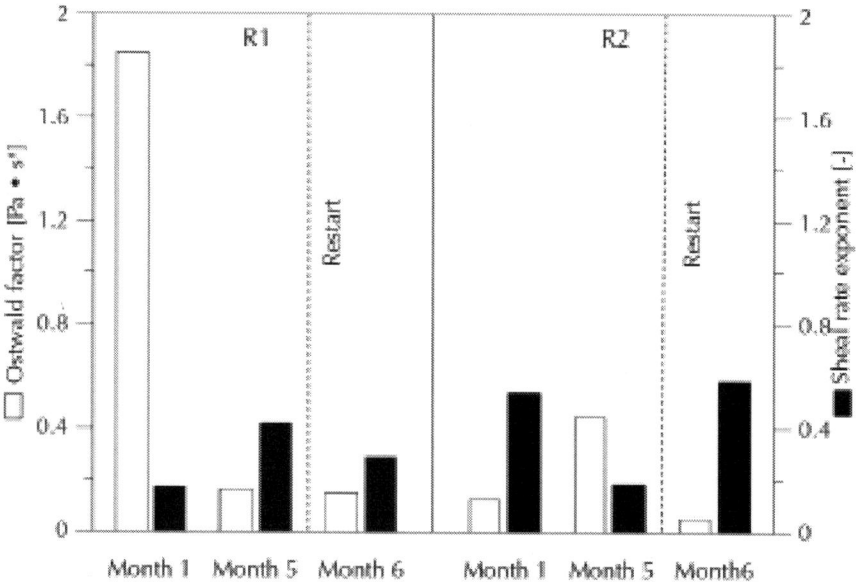

Figure 5: Ostwald factor and the shear rate exponent for R1 and R2. Ostwald factor and the shear rate exponent for R1 and R2 over 6 months covering the period of disturbed stirring. No considerable differences were detected during the stirring breakdown.

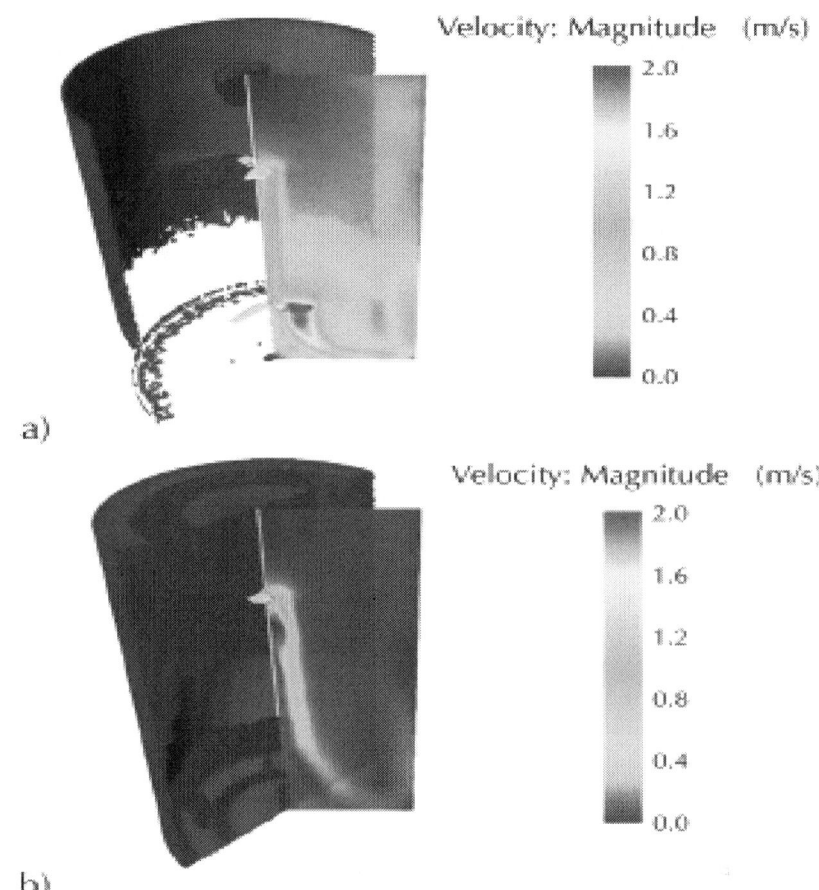

Figure 6: CFD simulations of the flow regime in the biogas plant with the rheology of the original substrate. (a) During normal operation. (b) With one broken paddle. The methanogenic reactor showed a good performance during normal operation; whereas the flow regime broke down and stagnant zones appeared when the reactor was only partially stirred.

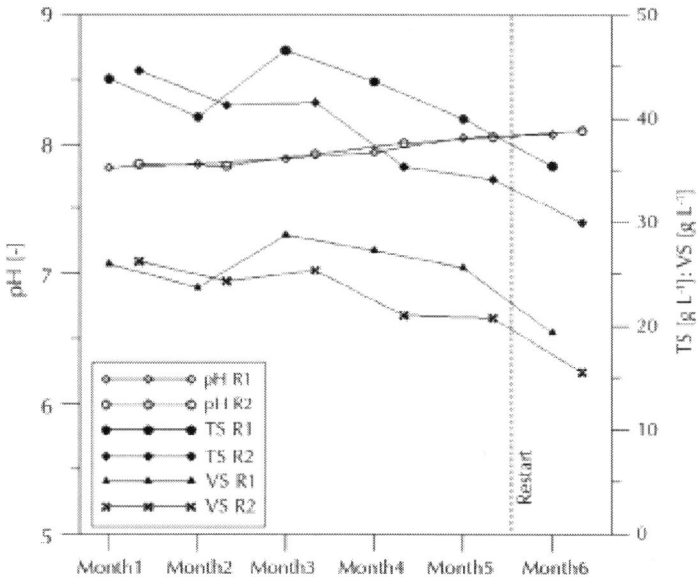

Figure 7: Average values of pH, total solids, and volatile solids. Lower TS and VS values were detected in R2. The pH value increased and TS and VS values decreased slightly after restart of the agitator in month 5.

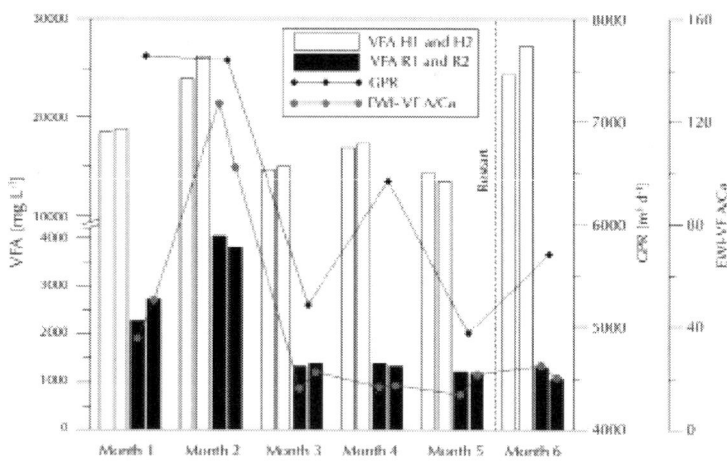

Figure 8: VFA concentrations, GPR, and early warning indicator (EWI-VFA/Ca). Highest VFA values in the hydrolytic reactors H1 and H2 in month 2 and month 6. VFA in the methanogenic reactors R1 and R2 increased and biogas

production decreased after agitator disturbance. Afterward, the concentrations of VFA decreased and stabilized.

Process Disruption-Associated Differences in the Microbial Community

Genetic fingerprinting of the bacterial community composition in the two methanogenic reactors showed a diverse band pattern after agitator breakdown (Figure 9 a,b). One month after the stirring was disturbed in both reactors, the microbial composition changed and differed between the two reactors. A characteristic band pattern for each reactor was visible at the genetic fingerprinting whereby mostly bacteria of the phyla Firmicutes and Bacteroidetes dominated the biocenosis (Table 3). Within the phylum Firmicutes, members of the order *Clostridiales* were dominant; whereas a *Proteiniphilum*-assigned organism from the phylum Bacteroidetes showed strong band intensities throughout the monitoring period, indicating a codominance in the reactors (band 19). In addition, one sequence was affiliated to a bacterium from the phylum Chloroflexi. Cardinali-Rezende et al. [34] as well as Leven et al. [35] investigated the microbial community of anaerobic reactors treating household waste and also observed bacteria of the Firmicutes, Bacteroidetes, and Chloroflexi as the dominating phyla. Although the band patterns of both reactors were similar in month 1, differences in the intensities were detected for several bands. A higher abundance of an unclassified bacterium was indicated by the more intensive band 7 in R2 (Table 3). From month 2 on, the band patterns of R1 and R2 differed significantly. The intensity of band 5, which was affiliated to a bacterium from the order *Bacteroidales*, became stronger in R1 and weaker in R2. The unclassified bacterium (band 7) was less dominant in R1 from month four to month five; whereas its dominance increased in R2. After the restart of the agitator in month 5, the band intensities from the *Bacteroidales* bacterium and the unclassified bacterium increased in both reactors again. The Pareto-Lorenz distribution pattern of R1 showed no differences in the functional organization of the microbial community during the agitator breakdown and afterward (Figure 10). Twenty percent of the cumulative number of bands was covered by 45% of the cumulative band intensities. By contrast, the Pareto-Lorenz distribution pattern of the partially stirred reactor R2 showed a difference in the functional organization of the microbial community in month 5 compared to months 1 and 6 as well as to the community in R1. In month 5, only 20% of the bands covered nearly 80% of the band intensities, indicating an uneven microbial community composition with few dominant species. In month 6, after restart of the agitator, the evenness of the microbial composition improved to a value of about 45% and was again on the same level as observed for R1 indicating a similar microbial community composition. Both reactors were affected by foaming in the first month after agitator disruption and additionally a floating layer formation was found in R2 when the roof of the reactor was opened to repair the broken stirring paddle. Foaming and bulking caused by filamentous bacteria is well known in wastewater treatment plants (WWTP) and anaerobic digesters treating activated sludge [7]. A *Proteiniphilum*-like bacterium from the phylum Bacteroidetes dominated the microbial community in both reactors. Filamentous members of the Bacteroidetes phylum have been isolated from many environments [36-39]. However, little is known about the involvement of these microorganisms in bulking or foaming up until now [40], and the morphology of species from the genus *Proteiniphilum* was described as rod-shaped [41]. Additionally, a member from the phylum Chloroflexi was identified in the reactors. Some members of this phylum have a filamentous morphology occasionally triggering the formation of foam and floating layers in WWTP[42,43]. Accordingly, microscopic analyses (Figure 11) revealed filaments in the floating layer. It is arguable if the filamentous morphology of the microorganisms was an adaption to the deteriorated distribution of nutrients due to the disturbed mixing in both reactors because the filamentous structure allows an improved nutrient absorption or whether the microorganisms were obligatory filamentous. However, it is not clear if the filamentous bacteria promoted the floating layer formation in this case. Most likely, they were just accompanying bacteria and the floating layer formation was mainly caused by fibrous substrates and especially the stirring failure. Furthermore, it has to be discussed if the alternating substrate mix might have had an influence on the microbial community composition in the reactors. Since the exact amount of introduced substrates for every month is not known by the authors, a correlation between substrate loading and microbial community change was not possible. However, both methanogenic reactors were fed by the same substrate mix with similar VFA as well as TS and VS concentrations.

Moreover, the microbial community structure only changed after the stirring differed in the reactors and adapted again after the stirring was restarted. Therefore, it is reasonable to regard the disturbance of stirring as the main trigger of the microbial community composition change. The band pattern of the floating layer showed no differences to the associated digestate (Figure 9c). Apparently, the microorganisms in the floating layer were not able to degrade efficiently the accumulated organic acids, although *Syntrophomonas*-like organisms which are known to withstand high VFA concentrations were detected in the floating layer [44]. Probably, the VFA concentrations were too high for a sufficient degradation leading to an inhibition of the microorganisms. Moreover, the accumulated fibrous substrates in the floating layer were less degradable, as also described by Heiske et al. [45].

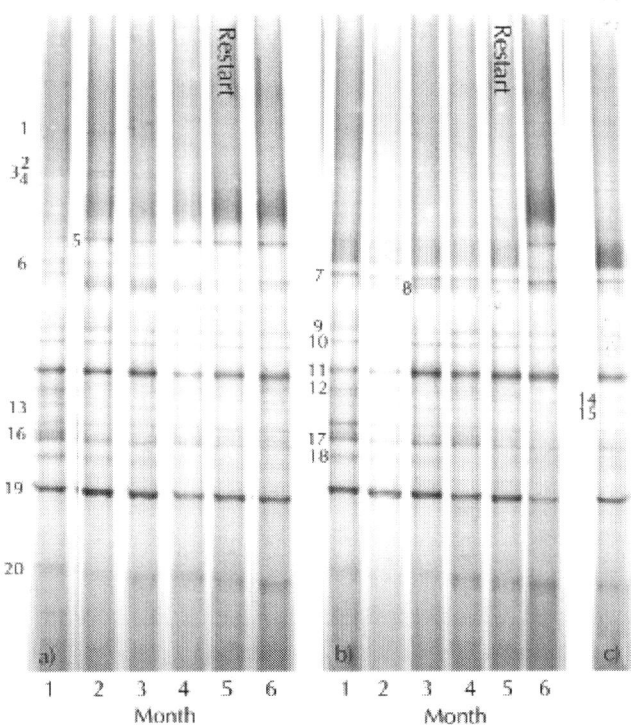

Figure 9: DGGE fingerprinting over 6-month monitoring. (a) R1 and (b) R2. Numbered bands were further identified by sequencing (Table 3). The bacterial community composition changed after the agitator disruption in month

1 and adjusted after restart of the agitator in month 5. **(c)** DGGE pattern of floating layer showed no differences to the associated digestate of R2 in month 5.

Table 3: Partial 16S rRNA gene sequences retrieved from DGGE fingerprint and sequencing excised bands

Band ID	Genbank accession number	Closest relative (accession number)	BLAST similarity (%)	Taxonomic classification
1	KF147561	Uncultured Clostridiales bacterium (JQ741982.1)	91	Unclassified Clostridiales
2	KF147562	Anaerobic bacterium (AY756145.2)	91	Unclassified Clostridiales
3	KF147563	Uncultured bacterium (GQ132906.1)	83	Unclassified bacteria
4	KF147564	Uncultured bacterium (JX224468.1)	83	Unclassified bacteria
5	KF147565	Uncultured bacterium (KC605949.1)	82	Unclassified Bacteroidales
6	KF147566	Uncultured Bacteroidetes bacterium (JX102011.1)	99	Unclassified Flavobacteriaceae
7	KF147567	Uncultured candidate division WWE1 (JX102010.1)	94	Unclassified bacteria
8	KF147568	Peptostreptococcaceae bacterium (AB377177.1)	94	Gallicola
9	KF147569	Clostridium sp. (FJ424481.1)	98	Clostridium sensustricto
10	KF147570	Uncultured bacterium (AB273806.1)	87	Unclassified Clostridiales
11	KF147571	Uncultured bacterium (AB850144.1)	92	Unclassified Clostridiales
12	KF147572	Uncultured compost bacterium (FN667344.1)	90	Unclassified Clostridiales
13	KF147573	Uncultured Firmicutes bacterium (FN429799.1)	90	Unclassified Clostridiales
14	KF147574	Uncultured bacterium (HQ156179.1)	91	Unclassified Clostridiales
15	KF147575	Uncultured bacterium (FN993992.1)	89	Unclassified Clostridiales

16	KF147576	Uncultured Syntrophomonas sp. (KF511597.1)	95	Syntrophomonas
17	KF147577	Uncultured Clostridia bacterium (JN998166.1)	91	Syntrophomonas
18	KF147580	Uncultured Firmicutes bacterium (HM041937.1)	92	Unclassified Clostridiales
19	KF147578	Uncultured bacterium (GQ134523.1)	88	Proteiniphilum
20	KF147579	Uncultured Chloroflexi bacterium (CU923171.1)	99	Levilinea

Taxonomic assignment was done by RDP Classifier with a confidence threshold of 50%. Closest relatives are shown including Genbank accession numbers.

Lienen *et al.*

Lienen *et al. Energy, Sustainability and Society* 2013 **3**:20, doi:10.1186/2192-0567-3-20

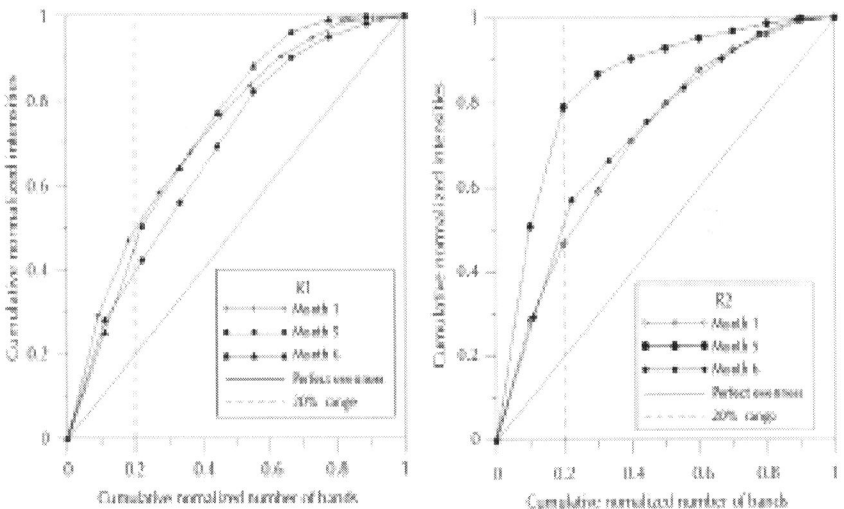

Figure 10: Pareto-Lorenz distribution pattern of R1 and R2 in months 1, 5, and 6. Perfect evenness is illustrated by straight line and 20% range is illustrated by dashed vertical line. Broken stirring paddle in R2 had greater influence on microbial evenness than complete downtime of agitator in R1.

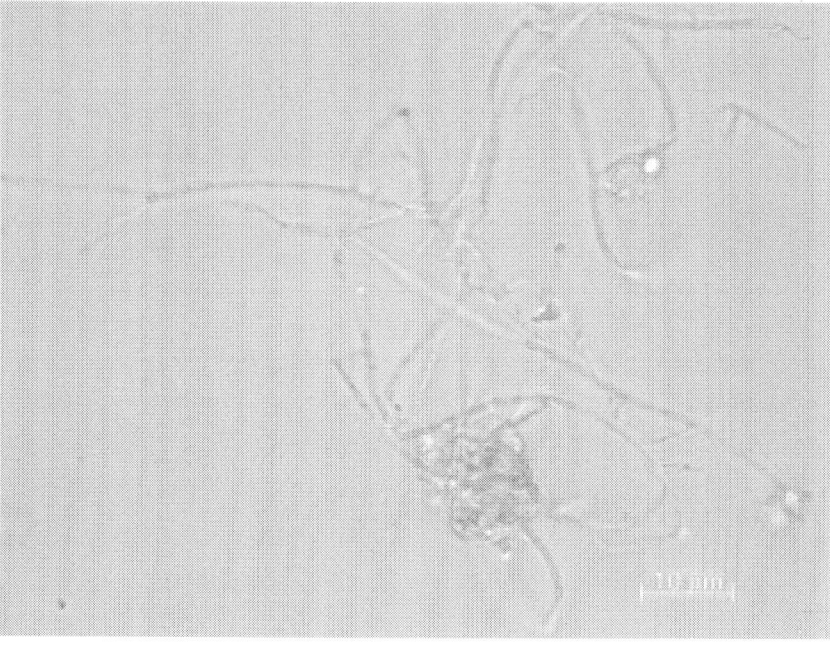

Figure 11: Filamentous structures in the floating layer sample (×100 magnification).

CONCLUSIONS

This study shows that the formation of floating layers and foam in full-scale biogas plants can be reduced by proper stirring. A precise knowledge of the rheology of the substrate mix and an appropriate agitation technology is essential for efficient mixing as well as optimized energy consumption. Furthermore, the stirring has a crucial influence on the microbial community structure. As most of the microorganisms are still uncultured and uncharacterized, further investigation on the microbial community composition is fundamental to enhance the efficiency of anaerobic digesters. Additionally, controlled substrate feeding as well as monitoring of EWI might prevent foaming due to overloading.

AUTHORS' CONTRIBUTIONS

TL carried out the molecular biological studies, sequence alignment, and drafted the manuscript. AK coordinated the sampling and carried out chemical analyses. MB conducted the rheological analyses and participated in the design of the study. LM and KG participated in the design of the study and performed detailed chemical analyses of the foaming event. MK and HW conceived of the study, participated in its design and coordination, and helped to interpret the results and to draft the manuscript. All authors read and approved the final manuscript.

ACKNOWLEDGEMENTS

The Authors wish to thank the German 'Federal Ministry for the Environment, Nature and Nuclear Safety' for funding the project 'Optgas' (FKZ 03 KB018) and the biogas plant operator for the cooperation. Special thanks go to Prof. Dr. Andreas Zehnsdorf and Dr. Roland Müller for helpful discussion and critical reading of the manuscript.

REFERENCES

1. Balussou D, Kleyböcker A, McKenna R, Möst D, Fichtner W (2012) An economic analysis of three operational co-digestion biogas plants in Germany. Waste Biomass Valor 3(1):23-41

2. Kleyböcker A, Liebrich M, Kasina M, Kraume M, Wittmaier M, Würdemann H (2012) Comparison of different procedures to stabilize biogas formation after process failure in a thermophilic waste digestion system: influence of aggregate formation on process stability. Waste Manag 32(6):1122-30

3. Kleyböcker A, Liebrich M, Verstraete W, Kraume M, Würdemann H (2012) Early warning indicators for process failure due to organic overloading by rapeseed oil in one-stage continuously stirred tank reactor, sewage sludge and waste digesters. Bioresour Technol 123:534-41

4. Moeller L, Goersch K, Neuhaus J, Zehnsdorf A, Mueller RA (2012) Comparative review of foam formation in biogas plants and ruminant bloat. Energy Sustainability Soc 2:12

5. Bischofsberger W, Dichtl N, Rosenwinkel K, Seyfried C, Böhnke B (2005) Anaerobtechnik. Springer-Verlag, Berlin Heidelberg.

6. Deublein D, Steinhauser A (2008) Biogas from waste and renewable resources. Wiley-VCH Verlag GmbH & Co. Kga A, Weinheim.

7. Martins A, Pagilla K, Heijnen J, van Loosdrencht M (2004) Filamentous bulking sludge-a critical review. Water Res 38(4):793-817

8. Westlund AB, Hagland E, Rothman M (1998) Foaming in anaerobic digesters caused by*Microthrix parvicella*. Water Sci Technol 37:51-55

9. Frigon D, Guthrie RM, Bachman GT, Royer J, Bailey B, Raskin L (2006) Long-term analysis of a full-scale activated sludge wastewater treatment system exhibiting seasonal biological foaming. Water Res 40(5):990-1008

10. Ganidi N, Tyrrel S, Cartmell E (2009) Anaerobic digestion foaming causes-a review. Bioresour Technol 100(23):5546-54

11. Barber W (2005) Anaerobic digester foaming: causes and solutions. Water 21 7 1:45-49

12. Pagilla KR, Craney KC, Kido WH (1997) Causes and effects of foaming in anaerobic sludge digesters. Water Sci Technol 36(6–7):463-470

13. DIN 38409–1 (1987) German standard methods for the examination of water, waste water and sludge; parameters characterizing effects and substances (group H); determination of total dry residue, filtrate dry residue and residue on ignition (H 1). Deutsches Institut Für Normung EV DIN.38409–1:1987–01

14. Dumas JB (1831) Procedes de l'analyseorganique. Ann ChimPhys 247:198-213

15. Eder B, Schulz H (2007) Anlagentechnik. In: Eder B, Schulz H (eds) Biogas praxis, 4th edn. Ökobuch Verlag. pp 70-118

16. BurchardCH, GrocheD, ZerresHP(2001)ATVHandbucheinfacher Messungen und Untersuchungen auf Klärwerken. Hirthammer Verlag München.

17. Metzner AB, Otto RE (1957) Agitation of non-Newtonian fluids. AIChE J 3(1):3-10

18. Wilkinson WL (1960) Non-Newtonian fluids, fluid mechanics, mixing and heat transfers. Pergamon Press, London.

19. Muyzer G, de Waal EC, Uitterlinden AG (1993) Profiling of complex microbial populations by denaturing gradient gel electrophoresis analysis of polymerase chain reaction-amplified genes coding for 16S rRNA. Appl Environ Microbiol 59:695-700

20. Amann RI, Stromley J, Devereux R, Key R, Stahl DA (1992) Molecular and microscopic identification of sulfate-reducing bacteria in multispecies biofilms. Appl Environ Microbiol 58:614-623

21. Hall TA (1999) BioEdit: A user-friendly biological sequence alignment editor and analysis program for Windows 95/98/NT. Nucleic Acids Symp Ser 41:95-98

22. Altschul SF, Gish W, Miller W, Myers EW, Lipman DJJ (1990) Basic local alignment search tool. Molec Biol 215:403-410

23. Wang Q, Garrity GM, Tiedje JM, Cole JR (2007) Naive Bayesian classifier for rapid assignment of rRNA sequences into the new bacterial taxonomy. Appl Environ Microbiol 73(16):5261-7

24. Lorenz MO (1905) Methods of measuring concentration of wealth. J Am Stat Assoc 9:209-219

25. Wittebolle L, Vervaeren H, Verstraete W, Boon N (2008) Quantifying community dynamics of nitrifiers in functionally stable reactors. Appl Environ Microbiol 74(1):286-293

26. Switzenbaum MS, Giraldo-Gomez E, Hickey RF (1990) Monitoring of the anaerobic methane fermentation process. Enzyme Microb Technol 12:22-730

27. Moeller L, Goersch K, Mueller RA, Zehnsdorf A (2012) Formation and suppression of foam in biogas plants – practical experiences. Agric Eng (Landtechnik) 67(2):110-113

28. Scholwin F, Liebetrau J, Edelmann W, Kaltschmitt M, Hartmann H, Hofbauer H (2009) Biogaserzeugung und -nutzung. In: Energieaus Biomasse. Springer-Verlag, Berlin Heidelberg. pp 851-990

29. Miltimore JE, McArthur JM, Mason JL, Ashby DL (1970) Bloat investigations. The threshold fraction 1 (18S) protein concentration

for bloat and relationships between bloat and lipid, tannin, Ca, Mg, Ni and Zn concentrations in alfalfa. Can J Anim Sci 50:61-68

30.   Stroot P, McMahon K, Mackie R, Raskin L (2001) Anaerobic codigestion of municipal solid waste and biosolids under various mixing conditions – I. Digester performance. Water Res 35(7):1804-1816

31.   Gomez X, Cuetos M, Cara J, Moran A, Garcia A (2006) Anaerobic co-digestion of primary sludge and fruit and vegetable fraction of the municipal solid wastes: conditions for mixing and evaluation of the organic loading rate. Renew Energy 31:2017-2024

32.   Kaparaju P, Buendia I, Ellegaard L, Andelidaki I (2008) Effects of mixing on methane production during thermophilic anaerobic digestion of manure: lab-scale and pilot-scale studies. Bioresour Technol 99(11):4919-4928

33.   Rojas C, Fang S, Uhlenhut F, Borchert A, Stein I, Schlaak M (2010) Stirring and biomass starter influences the anaerobic digestion of different substrates for biogas production. Eng Life Sci 10(4):339-347

34.   Cardinali-Rezende J, Debarry RD, Colturato LFDB, Carneiro EV, Chartone-Souza E, Nascimento AMA (2009) Molecular identification and dynamics of microbial communities in reactor treating organic household waste. Appl Microbiol Biotechnol 84:777-789

35.   Levén L, Eriksson AR, Schnürer A (2007) Effect of process temperature on bacterial and archaeal communities in two methanogenic bioreactors treating organic household waste. FEMS Microbiol Ecol 59(3):683-93

36.   Williams TM, Unz RF (1985) Isolation and characterization of filamentous bacteria present in bulking activated sludge. Appl Microbiol Biotechnol 22:273-282

37.   Kaempfer P, Weltin D, Hoffmeister D, Dott W (1995) Growth requirements of filamentous bacteria isolated from bulking and scumming sludge. Water Res 29:1585-1588

38.   Eilers H, Pernthaler J, Peplies J, Glockner FO, Gerdts G, Amann R (2001) Isolation of novel pelagic bacteria from the German bight and their seasonal contributions to surface picoplankton. Appl Environ Microbiol 67:5134-5142

39. Lydell C, Dowell L, Sikaroodi M, Gillevet P, Emerson D (2004) A population survey of members of the phylum Bacteroidetes isolated from salt marsh sediments along the East Coast of the United States. Microb Ecol 48:263-273

40. Kragelund C, Levantesi C, Borger A, Thelen K, Eikelboom D, Tandoi V, Kong Y, Krooneman J, Larsen P, Thomsen TR, Nielsen PH (2008) Identity, abundance and ecophysiology of filamentous bacteria belonging to the Bacteroidetes present in activated sludge plants. Microbiol 154(3):886-94

41. Chen S, Dong X (2005) *Proteiniphilum acetatigenes* gen. nov., sp. nov., from a UASB reactor treating brewery wastewater. Int J Syst Evol Microbiol 55(6):2257-61

42. Kragelund C, Levantesi C, Borger A, Thelen K, Eikelboom D, Tandoi V, Kong Y, van der Waarde J, Krooneman J, Rossetti S, Thomsen TR, Nielsen PH (2006) Identity, abundance and ecophysiology of filamentous Chloroflexi species present in activated sludge treatment plants. FEMS Microbiol Ecol 59(3):671-82

43. Nielsen PH, Kragelund C, Seviour RJ, Nielsen JL (2009) Identity and ecophysiology of filamentous bacteria in activated sludge. FEMS Microbiol Rev 33(6):969-98

44. Sousa DZ, Pereira MA, Smidt H, Stams AJM, Alves MM (2007) Molecular assessment of complex microbial communities degrading long chain fatty acids in methanogenic bioreactors. FEMS Microbiol Ecol 60(2):252-65

45. Heiske S, Schultz-Jensen N, Leipold F, Schmidt JE (2013) Improving anaerobic digestion of wheat straw by plasma-assisted pretreatment. J Atomic Mol Phys. Article ID 791353

Citations

CHAPTER 1

Wuchang Wang, Yuxing Li, Haihong Liu, and Pengfei Zhao, "Study of Agglomeration Characteristics of Hydrate Particles in Oil/Gas Pipelines," Advances in Mechanical Engineering, Article ID 457050, in press.

CHAPTER 2

Enbin Liu, Changjun Li, and Yi Yang, "Optimal Energy Consumption Analysis of Natural Gas Pipeline,"The Scientific World Journal, vol. 2014, Article ID 506138, 8 pages, 2014. doi:10.1155/2014/506138.

CHAPTER 3

Alex W. Dawotola, P. H. A. J. M. van Gelder, and J. K. Vrijling, "Decision Analysis Framework for Risk Management of Crude Oil Pipeline System," Advances in Decision Sciences, vol. 2011, Article ID 456824, 17 pages, 2011, doi:10.1155/2011/456824.

CHAPTER 4

Guofeng Du, Qingzhao Kong, Timothy Lai, and Gangbing Song, "Feasibility Study on Crack Detection of Pipelines Using Piezoceramic Transducers," International Journal of Distributed Sensor Networks, vol. 2013, Article ID 631715, 7 pages, 2013. doi:10.1155/2013/631715.

CHAPTER 5

Lekan Taofeek Popoola, Alhaji Shehu Grema, Ganiyu Kayode Latinwo, Babagana Gutti, and Adebori Saheed Balogun, Corrosion problems during oil and gas production and its mitigation, doi:10.1186/2228-5547-4-35.

CHAPTER 6

Daria Gritsenko and Johanna Yliskylä-Peuralahti, Governing Shipping Externalities: Baltic Ports in the Process of SOx Emission Reduction, doi:10.1186/2212-9790-12-10

CHAPTER 7

Mahendra Yadav, Usha Sharma, and Premanand Yadav, Corrosion inhibitive properties of some new isatin derivatives on corrosion of N80 steel in 15% HCl, doi:10.1186/2228-5547-4-6.

CHAPTER 8

Tobias Lienen, Anne Kleyböcker, Manuel Brehmer, Matthias Kraume, Lucie Moeller, Kati Görsch, and Hilke Würdemann, Floating layer formation, foaming, and microbial community structure change in full-scale biogas plant due to disruption of mixing and substrate overloading, doi:10.1186/2192-0567-3-20.

Index